I'm the Bob and cathy's kid

Emotions - Love and Fury

I'm the Bob and Cathy's Kid

A journey of hope and inspiration told by those who know of her challenges and celebrate her strengths.

By "The Book Team" –
B.A.Batt, S.Doane, J.Huff, T.Hurwitz A.Karst

Foreword by Dr. Trevor A. Hurwitz, MBChB, MRCP(UK), FRCP(C), Neurology and Psychiatry UBC Hospital

Suite 300 - 990 Fort St
Victoria, BC, V8V 3K2
Canada

www.friesenpress.com

First Edition — 2018

ISBN
978-1-5255-1704-4 (Hardcover)
978-1-5255-1705-1 (Paperback)
978-1-5255-1706-8 (eBook)

1. BIOGRAPHY & AUTOBIOGRAPHY, MEDICAL

Distributed to the trade by The Ingram Book Company

PREFACE

A human story, which shines scientific light into the soul of human behaviour.

CONTENTS

AUTHORS NOTE:

The title of this book, *I'm the Bob and Cathy's Kid,* was chosen because when Suzie was a little girl, and was introduced to new people, she would say "Hi, My name is Suzie, I'm the Bob and Cathy's Kid!"

Somehow, over all these years, this introductory phrase of hers has stuck with those who worked with her as a child, and so it seemed a fitting and natural title for the book.

Please enjoy this read as you share in her life's journey.

To Our Darling Suzie

FOREWORD

All Heart
There was a little girl,
Who had a little curl,
Right in the middle of her forehead.
When she was good,
She was very, very good,
And when she was bad she was horrid.
—Henry Wadsworth Longfellow (1807–1882)

Suzanna's story is a story about our emotional inner selves and the illusion of a single self. Her story is about who we are and who we strive to be.

We are in fact many selves that in normal circumstances work seamlessly together. We possess an emotional self, shared with all living creatures that has been pressured to evolve and adapt to survive and reproduce in a world of greater social complexity but still governed by the will to dominance. And then there is a human-only emotional self, struggling to achieve self-mastery and live co-operatively where actions and consequences have values.

Her story can be most easily understood using the model of brain evolution proposed by MacLean (image). The MacLean triune (three-in-one) brain has a central reptilian core brain that recognizes our basic needs for survival and reproduction—air, water, food, and sex. This was followed by the limbic emotional brain. The limbic emotional brain galvanizes the body into action or retreat or paradoxically provokes protective immobility. The purpose of the limbic emotional brain is to bring about engagement with or disengagement from the lived-in world where inanimate and animate objects must provide for these basic needs. Finally, a cortical mantle brain was added. A late-comer to our development, the cortical mantle is needed to bring the basic reptilian-limbic brain under the control of its owner. The cortical mantle brain allows a finessed grasp of incoming information and then fine tunes our motor and behavioural responses to maximize benefit and minimize harm to ourselves. It is also the presumed source of our free will and moral self and allows for value-based nuanced behaviour that matches the view of ourselves as we go about our daily business.

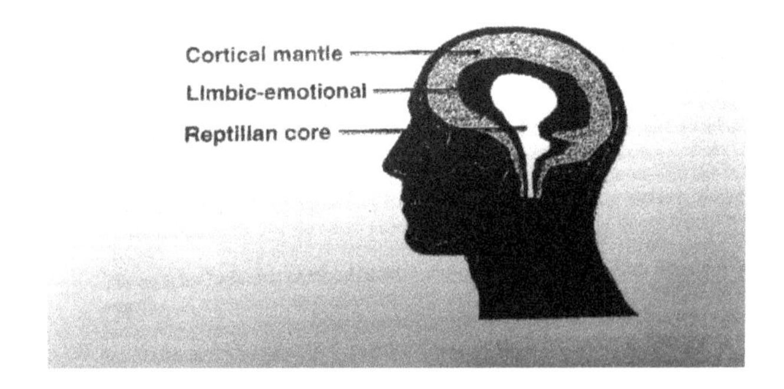

The cortical mantle brain is the most fickle and vulnerable to disruption and comes to finely graded variations within and between each individual, varying from moment to moment and from day to day. When we let down our guard, we are governed by our reptilian-limbic brain. Hyper strong emotions emerge, driving poorly regulated and adjudicated behaviours, expressed in full force with no consideration of consequences to self or others—all heart but revealing the beast within.

Suzanna has had one side of her brain's cortex[1] removed or disconnected from the rest of her brain in the desperate hope of controlling the epilepsy that was destroying her life. This right side was already malfunctioning because of her repeated seizures and the underlying nerve cell disturbances that were making these cells vulnerable to unregulated electrical storms. By removing this cortical mantle, she has lost all left-sided vision, and the left side of her body has lost sensation and is permanently paralyzed. Her neurological problems are left-sided because the right side of the brain controls the left side of the body and vice versa. Devastating and heartbreaking enough, but no one was prepared for the frightening consequences of the permanent loss of regulation over her inner reptilian-limbic core.

Life for Suzanna is now black or white, the beauty or the beast. In neuropsychiatric terminology, the most primitive strongest response activated by an internal need or an external stimulus is the one to surface. This response is driven and shaped by overpowering emotions, and the behaviours that follow reveal the original beast that lies within each of us. It is what you and I may experience and display when feeding alone (all fingers and shoveling food into our mouths) or driving anonymously (cursing out loud or gesturing rudely to the other morons on the road for their stupidity and selfishness).

Our cortical mantle brain gives us the ability to suppress these strong primate responses and then consider and select amongst the hierarchy of options the one that will get the job done best—the one that will deliver the maximum benefit and cause the least harm for and to ourselves and others. This is usually done unconsciously and ultra-rapidly. In our vulnerable moments as when fatigued and highly stressed, the reptilian-limbic brain takes charge and we act in the heat of the moment, only to regret our behaviour when we cool down and

1 Cortex: 1. The outer layer of grey matter that covers the surface of the cerebral hemisphere. 2. The outer layer of an organ or other structure as distinguished from its inner substance called medulla.

the moment has lost its passion. Nothing to be proud about and not a pretty picture, embarrassingly at variance with the elevated ethereal beliefs that we hold about ourselves.

Those who lack this cortical mantle are permanently all heart. They are intensely emotional without any breaks to hold them back. They may regret their actions later, but asking them to exercise control or learn to impose control from repeated feedback and the negative consequences of their behaviour is like trying to slow or stop a train in motion that has no brakes. The knowledge of cause and effect, of actions and their consequences, is there, but the machinery to hold back to allow a more nuanced response is missing. Knowledge cannot and does not govern behaviour. It is bulldozed aside by the more powerful reptilian-limbic brain, and a Suzanna is the result. This is a well-known problem in neuropsychiatry. We call it the disinhibited frontal lobe syndrome with loss of inhibitory control, and it happens when the front part of the brain is injured. Suzanna has lost the entire right cortical mantle, and the loss of the front part is responsible for her behavioural difficulties.

When in the moment, Suzanna is all heart. To expect otherwise and despite years of feedback, debriefing, and education is to completely misunderstand the challenge that she and her caregivers must face on a day-to-day basis. She is and will always be in the moment for every moment of her waking day. She is and will always be all love and all fury. Affectionate, "you cutie patootie," and hugs and kisses revealing her true individualized personality—warm, kind, loving, and determined to do her best, or unconstrainedly aggressive and anxiously disorganized revealing her limbic-reptilian brain that is an undifferentiated beast that we all share in common with her.

Any dysfunctional behaviour is her emotionally hyper-charged inner needs clashing with the proprieties and external constraints of the shared, lived-in world. We are no different. All our needs and life encounters provoke an emotionally charged need-reaction pairing. This is often unconscious and passed by unnoticed yet still influencing our behaviour. If consciously experienced, our behaviour is brought under some degree of control, depending upon circumstances, our degree of insight into options and consequences, the intensity of the activated emotions, and our active willingness to curb our primitive, strongest, and first-to-arrive responses. But in Suzanna, the most primitive, strongest, and first-to-arrive response is always expressed in full force to the shock and dismay of others in her presence who risk moment-to-moment boundary violations and physical harm.

To help Suzanna reach her potential without physical restraints or medicinally-induced zombification requires the external world to anticipate the internal and external triggers that activate her needs, or respond to her disruptive behaviour in ways that go against our natural inclinations. The result is a set of strategies that brings out her individual self but also recognizes the early signs of a brewing storm. This requires constant vigilance coupled with softly spoken and carefully chosen words and actions all wrapped in genuinely felt caring and compassion. She needs a caregiving approach that involves social capitulation and placatory and flattering remarks that cede to our reflexive competitiveness when we are challenged or threatened. Suzanna may not be able to control her reptilian-limbic brain, but she still registers the world and its signals. Her caregivers must either anticipate and redirect internal and external triggers or otherwise respond with behaviours that soothe her impassioned mind. A monumental and exhausting task for which many are not equipped and has taken years of trial and error to develop. These care providing strategies will always be a work in progress as she moves through life phases and the ups and downs of normal daily living.

Who are these people who can do this? They are also all heart but of the highest kind, and that which separates us from all other living creatures. This is the emotion of morality. Morality is the emotion of goodness and badness, and like all other emotions, galvanizes actions or induces restraint consistent with its dictates. The locus of morality, together with consciousness, remains unknown to modern neuroscience, but for many reasons most likely resides in the cortical mantle brain. It has come late to living creatures, and this specific cortical capacity seemingly possessed only by one—us.

Consciousness and morality allow self-recognition and the ability to recognize other humans and their humanity—to see them as they might see themselves, an ability known as theory of mind. It allows us to pursue justice, love, goodness, and mercy and to walk humbly. It allows us to see through the frailties and vulnerabilities of others. It allows us to recognize that Suzanna's behaviour is not under her control, that she cannot be expected to slow down the train. This is not from an absence of morality, calculated malevolence, or disregard for or caring about others. She is kind, caring, and loving. It is because her brain has no brakes. Knowing this allows us to repeatedly forgive her offensive language and tolerate being cuffed, bitten, or spat at. So who are these people that are her family and her care providers? All saints, every one of them.

This is her story, this is their story. Unsanitized.

Dr. Trevor A. Hurwitz, MBChB, MRCP(UK), FRCP(C)
August 2017

INTRODUCTION

Perhaps one of the most important tasks facing every one of us on a daily basis is effective communication. Communication is vital to our sense of well-being. Enjoyment of social activities, success on the job, and warmth in personal relationships all depend on it. The ability to communicate is crucial to how we make our needs known and how we can meet the needs of others. Communication is the backbone of how we live our lives.

This is a story about Suzanna Bailey, who was born on August 4th, 1981 in Langley, BC. Suzie was born into a family that communicated love, joy, and affection ... a family that welcomed and nurtured her. She developed normally until, as a toddler, she suddenly began to experience seizures.[2] These seizures led to countless appointments

---

2 **Seizures and Epilepsy: A 2014 Revised Definition of Epilepsy/Epilepsy Foundation.** Seizures and epilepsy are not the same. An epileptic seizure is a transient occurrence of signs and/ or symptoms due to abnormal excessive or synchronous neuronal activity in the brain. Epilepsy is a disease characterized by an enduring predisposition to generate epileptic seizures and by the neurobiological, cognitive, psychological, and social consequences of this condition. Translation: a seizure is an event and epilepsy is the disease involving recurrent unprovoked seizures.
The human brain is the source of human epilepsy. Although the symptoms of a seizure may affect any part of the body, the electrical events that produce the symptoms occur in the brain. The location of that event, how it spreads, and how much of the brain is affected, and how long it lasts, all have profound effects. These factors determine the character of a seizure and its impact on the individual.
Having seizures and epilepsy also can affect one's safety, relationships, work, driving and so much more. How epilepsy is perceived or how people are treated often is a bigger problem that the seizures.

with medical professionals. Successive brain surgeries left her with life-changing brain injuries, resulting in physical disabilities and behavioural challenges.

What follows is an account of those who lived and worked with Suzie and their commitment to helping Suzie in her ability to communicate effectively. It is a journey of hope and inspiration told by those who know of her challenges and celebrate her strengths. It begins with her family and their experiences. There are general and special details of medical interventions. Dedicated professionals give accounts. There are descriptions of her education and behavioural strategies by those who believed in her ability to communicate despite her disabilities. As well, there is a record of the changes in her living accommodations and how those have evolved over the years.

What follows is meant to serve as a tool of encouragement. It demonstrates that perseverance and dedication can and do make a difference. It proves that any human being, despite seemingly insurmountable challenges, has the potential for effective communication. We hope that this story will touch the reader's heart and awaken compassion for the journey of parents who have children with medical challenges. If you are able to find even just one successful strategy or idea in your life journey as a parent or staff working with someone that has physical or mental challenges, then our goal has been achieved.

CHAPTER 1
Suzie's Family

Excerpt from an interview
WITH CATHY BAILEY, SUZIE'S MOTHER:

(...) Suzie was placed at Sunny Hill Health Centre for Children because of her post-hemispherectomy left side paralysis. She was at Sunny Hill for three or four months. It was a hard fit. We were just learning about how challenging Suzanna was really becoming. When children are little and suffering, everyone's heart goes out to them, but when they get older and bigger, that really changes. (...)

ABOUT SUZANNA
By Cathy Bailey, Suzanna's mother

When Suzie was born she was bigger than our other two girls. They were just over 6 lbs., and Suzie was 8 lbs. 15 oz. She was a very robust and beautiful baby. I remember that she was a happy baby, lovely, and even-tempered. She could entertain herself. You could put her on a blanket and she would just play. The only thing that was somewhat unusual was her crooked smile, which we thought was endearing.

One Sunday morning when she was fourteen months old, I noticed a different sound coming from Suzie's room—not alarming, but slightly different. We had decided to leave the kids with their grandparents, who were visiting for the weekend, while we went to church. When we returned, my father-in-law was in the driveway, madly waving his arms at us. He looked like he had aged ten years! We never even got out of the car; we rolled down the window to Grandpa and were told that while Suzie was eating breakfast something had happened, as if she were choking or something, and she had been taken to the hospital.

When we arrived at Langley Memorial Hospital, there were a number of medical personnel around her bed. They said to me, "Shhh ... she's going to have another one." I'd never seen a seizure before, and what I saw terrified me. I thought she was going to die. The seizures were about thirty seconds long. As parents, we were in shock.

Suzie had about six seizures that first day. She was put on a medication called *Phenobarb* in an attempt to control the seizures. The doctors asked me if she had had any vaccinations. Yes, she had an immunization within the previous week. Initially the doctors suspected that the vaccination may have been the cause of these seizures, however in later years, she had an MRI which showed dysplasia in the brain. When Suzie's cortex was developing in utero, the brain cells did not develop properly, and resulted in polimacrogyra (brain cells are larger and there are more of them, which causes the brain waves to have more valleys that normal.)

Suzie began to experience seizures every day while she was in the hospital. The medication they gave her did not really alleviate this problem, although a doctor told us that it would take a while for the medication to start working. We quickly came up to speed on learning how to watch for signs of seizures.

While Suzie was in Langley Memorial, Bob asked to have her transferred to Children's Hospital in Vancouver. A young Scottish doctor, Dr. Kevin Farrell, was new to the neurology department there. He examined Suzanna. Then Bob and I had a conversation with him. He said, "Don't let anyone kid you. You have a very sick kiddie here." He used the words "epilepsy" and "seizure disorder." Dr. Farrell suspected it was a focal kind of epilepsy, and that is what they pursued. He performed a complete neurological assessment, motor assessment, EEG and vision tests, and ordered extensive blood work. Subsequently, she had an assessment at the University of British Columbia (UBC), looking for a tumour possibility.

What happened from the time she was fourteen months to the time she was two and a half years old was a litany of medications and treatments, which is now a complete blur in my memory. You could smell *paraldehyde* everywhere. It was intended to stop the seizures and was administered in many different ways, including suppositories. Bob and I were frustrated that nothing was getting better; in fact, her seizures were getting worse. Much worse, more frequent, more kinds, more terrifying!

It seemed everything was escalating! She was in and out of Children's Hospital constantly. Bob and I were continually panicked, feeling stressed and worried. Thankfully, we had family members who helped with our two other children at home.

At one point, Suzie was admitted to the ICU to be put into a coma. The medical staff lowered all her bodily functions except her heartbeat. They could only keep her in that state for twenty-six hours, hoping that maybe they could jolt the brain out of the pattern it was in. We watched them do this as well as bring her out of the coma. It was very scary, and all we could do was pray for her. The EEG showed nothing in her brain. When her brain function came back slowly, we were hopeful, but the treatment did not work. The seizures came back more or less immediately.

Even in these difficult times, Suzanna's personality served her very well. She was plucky and exceedingly independent. She was accustomed to coping and she was not whiny. As parents, that made things easier for us. Her doctors, nurses, and therapists became so important to us and gave countless support and encouragement, despite the many questions plaguing their investigations.

As time went on, the seizures started to affect her development. She was struggling to cope and to understand things. We started to see her slowing down.

We eventually went to Montreal for Suzanna's first surgery when she was two and a half. Bob stayed home with our other two daughters but flew out for the final consultation and the surgery. He was with me from the end of November until mid-January.

The Montreal Neurological Institute was the only hospital in Canada that could perform the type of surgery recommended. It was there that the specialists began to shed light on an actual diagnosis. They discovered that there were multiple foci* throughout her brain, which were widespread in her cortex (cortical layer). *

All told, Suzanna benefited from her next four surgeries. She was in Montreal each time for about eight weeks. The seizures decreased from approximately seventy to eighty a day to about thirty. She had to wear a helmet all of the time, but was walking and running. Suzanna had been a beautifully formed child with long legs, but her physical appearance gradually changed due to her many medications and the side effects of surgery.

Many people were helpful in making the treatment possible. During one of the trips to Montreal, I left my purse on the plane. There was a severe snowstorm in Montreal at the time. A young man at the airport saw that I was upset and asked me what was wrong. I told him my problem: I had no money for a taxi and I had to get Suzanna to the hospital. That young man took care of everything for us. We were convinced he was an angel. He contacted a flight attendant to retrieve my purse and made sure we got safely to the hotel, where we stayed for three weeks.

As you can imagine, the costs of these trips to Montreal were becoming enormous. Our pastor helped us make contacts, and through him we met a lovely couple in Montreal. We stayed with them and are still friends today. I made another friend with whom I stayed and met amazing women at the hospital who became friends of mine.

There was also a Ronald McDonald House to support us where we stayed on several occasions.

In terms of support, so much credit also goes to Children's Hospital in Vancouver. It was extraordinary. At the time they had never had a child stay with them for the length of time Suzanna did on 3A (the "Neuro Ward").

By the time Suzanna reached ten years old, she continued to have intractable seizures. She was given massive amounts of *gamma globulin,* but that did not work.

Another surgery was planned in Montreal, and as parents we found this prospect extremely difficult. We prayed that we were making the right decision for Suzanna's future. We did not know if this operation would indeed end the seizures, and we had no guarantee of a totally positive outcome. We struggled with the fear of a changed person given that the problem was a malfunctioning brain. We worried each time that our dear daughter wouldn't "come back to us."

This fifth surgery was the most extensive. A hemispherectomy was performed, and her left side became paralyzed, similar to the effects of a stroke. This affected her gait, her arm, face, and eye. (More on this surgery in Appendix A.)

After the hemispherectomy, Suzanna was taken to Children's Hospital in Vancouver to be medically stabilized. We were told at that time there was nothing more that could be done.

From Children's Hospital, Suzanna went to Sunny Hill Health Centre for Children in Vancouver. The doctors wanted her to receive intensive physiotherapy. There was a point in time when it became very difficult for the nurses and hospital staff to manage Suzanna's behavioural issues. The focus for the nurses was on rehabilitation of medical issues, and behavioural interventions were not what the hospital was staffed to address.

Suzanna was confused, and her behaviour demonstrated that. She was locked in a safe room behind glass. She got out of an isolation crib that had plastic around the sides. She flooded the room. They had her on a mattress because she destroyed the bed. Her sisters would take her for walks across the ward with intravenous tubes trailing behind her. She would write on the walls. We brought cleaning supplies for her to clean up the walls in order to have a consequence for this behaviour. We struggled to parent as well as comfort our child. A note was written by Dr. Farrell to "prescribe" that Suzanna clean up the mess as therapy, as cleaning the walls was not a task allowed for other than unionized hospital staff.

Suzanna was at Sunny Hill for three to four months. It was a hard fit. We were beginning to learn about how challenging Suzanna was becoming. When children are little and suffering, everyone's heart goes out to them, but when older, they are expected to be more responsible. People's sympathies really change. We learned to advocate on our daughter's behalf, given her growing problems.

Even with the challenges Suzanna presented, the medical professionals did so much for us. I remember a conversation with a nurse. "I just can't get over how the staff has done so much for us," I said.

She said, "We just can't do enough for people like you. You never complain about the wait and are always thankful."

We knew we had a long road ahead of us, but we also knew we had to work with people and show gratitude. People were so well deserving of it.

Dr. Farrell said, "If there is anything we can do, just say it and we'll make it happen."

Dr. Farrell met with Bob and me for a compassionate discussion on how to deal with Suzanna's behavioural issues. It was decided that Suzanna could not go back to our home at that time. She needed rehabilitation and physiotherapy on a daily basis. Her behaviour was too difficult for us as parents to deal with on a full-time basis. Bob and I believed strongly that we had to somehow keep the threads of our family together, and we also agreed that we needed help to care for her.

The social worker at Children's Hospital called the Ministry of Children and Family Development and Suzanna's social worker. A meeting was arranged between ourselves and those who could offer assistance in planning this transition. This meeting included the local agency that provided service and support to families who had children with special needs in the Langley area. Dan Collins, the Executive Director of the Langley Association for the Handicapped (LAH), now the Langley Association for Community Living (LACL), was instrumental in navigating the funding sources as well as the resources needed to provide a new home in Langley for Suzanna.

The Ministry of Children and Family Development wanted to place Suzanna in a group home, and they were looking at the options province-wide, even as far away as Prince George. Bob and I insisted that Suzanna should stay in her community, and the ministry accommodated that request. That is when LACL came in to help. Suzanna became one of the highest-funded children in B.C. at the time, with double staffing at times. We were asked: "What do you want for Suzanna?" Bob and I immediately ruled out any option of Suzanna being placed in foster care, as she already had a family. We knew that nobody would love Suzanna like her family, but we knew that somebody could help us with care for her.

The current model for Suzanna's care was developed and proposed as one of the first of its kind. A small home was rented close to the family home. Staffing and guidelines for service were implemented. It was planned that Bob and I and our other children would have six months of respite while Suzanna was supported in this new home, visiting often without restrictions. We called the home "Benz" after the name of the street it was on. It was hoped that Suzanna could return home in six months, but this model served all of us well until she turned nineteen. It must be said that she distinguished between her family home and Benz. Over time, she understood that Benz provided a supportive place where she could live at a manageable pace and have her special needs met.

Life at Benz was up and down. While Suzanna excelled at learning things that the medical professionals did not think possible, the seizures continued. There were side effects of many drugs that were tried to control the different kinds of seizure activity. Children's Hospital and Dr. Farrell were still offering emotional support to Bob and me as well as to the staff of Benz. We were all baffled as to how to control these seizures and the behaviours that seemed to accompany them.

Finally, at fifteen, Suzanna had her sixth and last brain surgery at Children's Hospital in Vancouver. After this surgery, the surgeon said to the staff at Benz, "Do not let her hit her head." That night, she had a violent seizure, which caused her to come flying out of her bed and hit her head on the closet door. No damage was done, but it scared Bob and me as well as the staff, who were constantly concerned for her safety.

When Suzanna was about seventeen, her medical team wanted, as a last resort, to try a non-surgical approach to decreasing her seizures. It was hoped that this would also diminish the challenges with her behaviour. They installed a vagal nerve stimulator under the skin of her chest, which she has to this day. It was somewhat successful, temporarily, but has not achieved a lasting effect.

Suzanna's losses of impulse control and subsequent unpredictable behavioural outbursts at school and home have challenged us all. We spoke about this with Dr. Farrell, and he referred her to Dr. Hurwitz at the University of British Columbia. Dr. Hurwitz at UBC was considered a divergent thinker and a very compassionate person. He remains that way for us to this day.

It is difficult for us to remember the many details of our journey, trying to control Suzanna's seizures and behaviour. It was challenging to find the energy and the time to work, to parent all of our girls, attend the many medical appointments necessary for Suzanna, and keep our own lives together.

It was our deep faith that guided Bob and me through these very difficult years. For that words are inadequate, and for the many people whose love has supported us, we are grateful.

Home Alone
BY SYLVIA DOANE

Once a week, Suzie's mom, Cathy, would come for a visit and dinner at Suzie's house, giving the support worker time to go grocery shopping. While at Suzie's one evening, Bob called Cathy to see if she could pick him up. He was having some problems with his car. He was only five or six minutes away. Rather than get Suzie all dressed up on a very cold night, Cathy decided Suzie would be fine staying alone for a few minutes, snuggling in her favourite chair, colouring and watching TV. Cathy asked Suzie if she would be okay staying alone, and Suzie said, "Oh sure, I can take care of the house until you come back." Suzie had never been left alone before, but moms can make that decision, after all. Who would know? Right?

As soon as Cathy left, Suzie picked up the telephone and called her support worker, Sylvia, at home and let her know she was all by herself and looking after the house. Sylvia kept Suzie talking on the phone until Mom walked in the door. Cathy said she nearly died seeing Suzie chatting on the phone. She asked who she was talking to, and Suzie said she had called Sylvia to let her know she was alone and looking after the house while Mom went out to pick up her dad.

Sometimes a person just can't get away with a thing!

Suzanna Diane Monique Bailey

"I hate this helmet!"

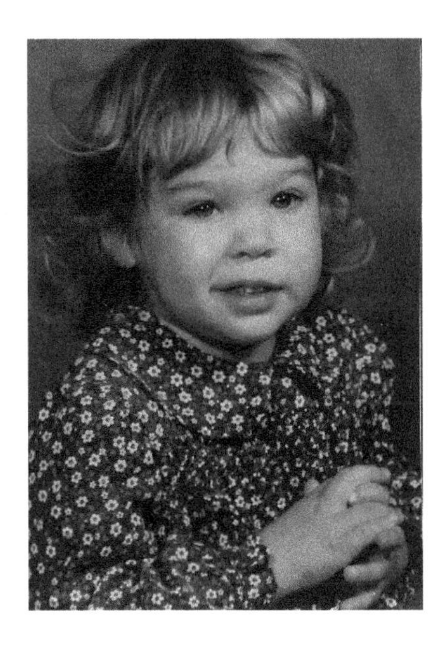

"What's happening to me?"

"Hanging out with my sister"

All of Daddy's girls

"Don't tell me what to do!"

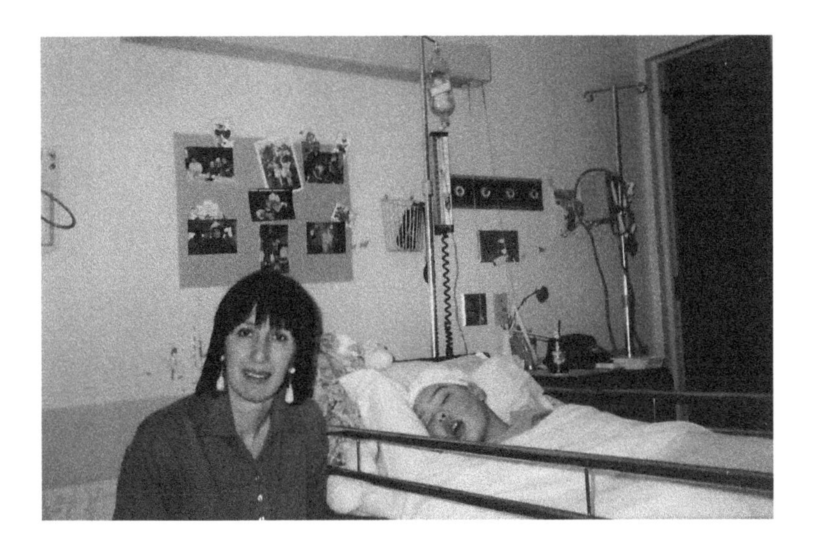

"Some of my most frightening times."

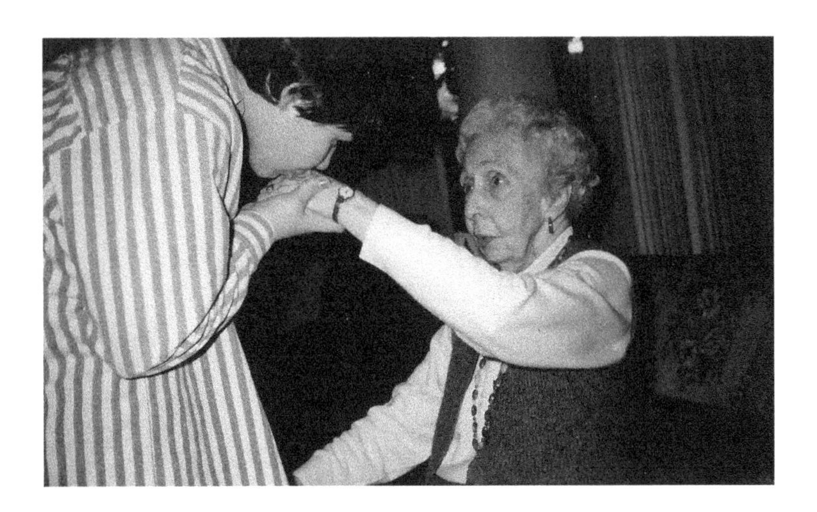

"Her majesty, my grandma!"

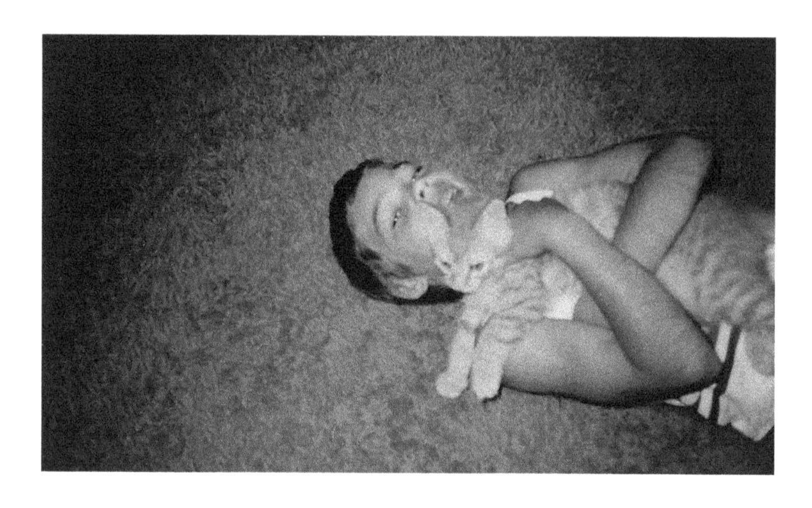

"I love cuddling my cat!"

"I'm gonna be a star!"

An excerpt from an interview WITH BOB BAILEY, SUZIE'S DAD:

I remember her trying to open the door with her left hand. She asked me, "Dad, what's wrong with my damn arm?" I don't know how we can live with ourselves knowing that we have crippled our own child. I don't know if we could have done it if we were faced with the same decision again. But we had no choice. The doctor said that if she continued with over two hundred seizures a day, she was literally going to "fry" her brain. We didn't know how to handle everything, other than just continue to pray about it. We asked why, but we never came up with an answer other than: "This is the path you have to walk."

INTERVIEW WITH BOB BAILEY, SUZANNA'S FATHER

By "The Book Team"

Did you ever wish Suzanna was born without the challenges that came with her?

I never thought of that, but if you ask me today, I would never wish for her to have to go through all of this again. I have memories of Suzie knocking on the neighbour's door, asking whether she could play with her friends, and being rejected. One time there were two kids who wrote some nasty words on Suzie's bicycle. I went to confront the dad, but he denied that his kid had done it. The mother of the other boy, who did do it, confronted her son and dealt with it. She offered to clean the bike or replace it. The boy admitted to what he had done, cleaned the bike, and offered Suzie $5.00 from his piggy bank. Suzie immediately forgave him and offered to go to the movies with him. She was about five years old at that time.

I don't think we could have gone through everything we faced without help from Dr. Farrell. After the first surgery, I was literally numb for two days. You could have cut off my arm and I wouldn't have felt it! We had lunch following the surgery, and I remember the food being completely tasteless.

When Suzie came out from the hemispherectomy, we knew she would never be the same physically, but we didn't know the degree to which she would be paralyzed. I remember her trying to open the door with her left hand. She asked me, "Dad, what's wrong with my damn arm?"

I don't know how we can live with ourselves knowing that we have crippled our own child. I don't know if we would have done it if we were faced with the same decision again. But we had no choice. The doctor said that if she continued with hundreds of seizures a day, she was literally going to "fry" her brain.

We didn't know how to handle everything other than just continue to pray about it. We asked why, but we never came up with an answer other than: "This is the path you have to walk."

One summer when the girls were still little we went camping on Vancouver Island. Suzie was climbing on everything. We had to watch her because she could have fallen into the fire. We planned to go for three weeks, but we so desperately needed time away as a family that we camped for five days longer. I felt so inadequate because I couldn't make her better. I was desperately trying to pull my family together, but even during the trip we sometimes had to rush to the hospital in the middle of the night. The staff gave her medication called *Paraldehyde*[3]

3 **Paraldehyde (Oral route, Injection route, Rectal route).** Paraldehyde is used to treat certain convulsive disorders. It is a central nervous system depressant. Paraldehyde oxidized in air turning brown and producing an odour of acetic acid.

to treat the seizures. They gave me a shoe box full of medications. I did not want to have that many medications with us, so I asked for only two weeks' supply at a time.

You get by on very little sleep. The situation numbs you. You are constantly getting a sense that things aren't good. I paid a price, and my girls paid a price. I'm surprised I'm still married and that I still have a relationship with my daughters. There were always traumatic problems with Suzie, and they weren't getting solved.

One time I was in an important meeting with a lawyer. My pager went off. It was a message from the ICU. In the middle of the interview, I had to ask the lawyer whether I could use the phone. The voice at the other end of the line said: "You have to come right now. Your wife needs you." I explained my situation to the lawyer and he said, "It's obvious that things aren't good because you turned white when you got the call." This kind of stress is how we lived for ten years. We continued to get calls in the middle of the night for years.

I remember Suzie being rejected numerous times by her peers when she asked to play with them. If she had not been faced with those surgeries, she would have been a phenomenal athlete. She was an unstoppable climber.

Dr. Farrell once said that he had never had a child as resilient as Suzie. He gave me examples of other children around the world who had literally become comatose from the seizures.

She has always had "shiveries" (Suzie's name for this particular kind of seizure). Although she would have breaks from the focal seizures for a period of time, they would always return.

There was a time when at least twice a week I would have to grab her at 2:00 a.m. and rush to the hospital. One time there was a new doctor on duty when I brought her in. She was covered with bruises from falling down from the seizures. The doctor asked where all the bruises were from. The nurse on duty put herself between me and the doctor and said, "its okay, doctor, we know this family." Sometimes I brought her in when she was blue in the face. They told us that if she went into cardiac arrest, they didn't have the proper equipment or the right size of tubes to resuscitate her. She would have to be taken to Children's Hospital.

Suzie was incredibly persistent and sometimes embarrassingly so! She could get out of anything, any kind of safety bubble they put her in. She could find her way out of any restraints. She would get out while we were in the shower and she would go for a walk. Strangers would bring her back, asking, "Is this your child?" Somebody once saw her walking down the street to go and get some ice cream at Safeway. One of the neighbours once called and said, "She is in my fridge." She was on some kind of steroids that increased her appetite and caused her to gain twenty per cent of her body weight. In the hospital in Montreal, she would get fed first because she was pounding on the walls demanding, "I want my dinner!"

How do you think Suzanna's illness affected her sisters?

They had an absent father. Not that I wanted it that way, but I was trying to work, survive on very little sleep, pay the bills, and cope with worry, worry, worry all the time. I feel that our girls were shortchanged for those ten years. I don't remember going to any sporting events with them.

Once her eldest sister, Sara, was kneeling by Suzie's bed praying for her. She thought she had done something to cause the seizures in her sister. She carried a lot of guilt. Her other sister, Sam, was very protective of Suzie. I'm not certain of the full impact it had on them, but they felt very responsible and very protective of her. They knew that she wasn't purposefully trying to be difficult.

There was a point when they couldn't go to Nana's for the weekends because it was too much for Nana to handle both Suzie and her sisters at the same time.

From time to time we arranged for respite. Once, we found someone who reassured us she was very experienced with kids. Then she called at 2:00 a.m. and said, "Get this kid out of my house!"

I called Dr. Farrell when we couldn't handle it any more. He asked, "Is she having seizures?"

I said, "What are you talking about, Kevin? When did she not have seizures?"

He asked again, "Is she having seizures?"

I said, "Of course she is having seizures."

He replied, "Then take her to emergency."

When Suzie was only ten years old and Sunny Hill Health Centre for Children could no longer keep her, you and Cathy had to ask for help. Can you tell us how difficult that was for you?
It was painful and scary. It took me the longest time to understand why people would want to work with someone like Suzie and sincerely care for her. There was one time when Suzie was home with us for the weekend and she said, "I want to go home now." It was very sad but reassuring at the same time. Dr. Farrell explained to us that Suzie needed a very strict routine. It was important for her to know exactly what would happen and where to go. I think Suzie is better adjusted now than she would have been had she stayed home with us.

Help for Suzanna and her family came from Dan Collins, Executive Director of Langley Association for Community Living. It was important to the staff and management that the program was geared toward her needs and be very personal, planned specifically for her.

Pee in the Bushes

BY SYLVIA DOANE

Suzie's dad, Bob, drove Suzie home after having dinner at their home. Suzie walked into the kitchen and after a couple of minutes I asked where her dad was. Her answer was, "Having a pee in the bushes!"

MEMORIES FROM SAM (SUZIE'S OLDER SISTER)

When I think of my memories of being young and growing up with Suzie, I think of her as our sweet, fun playmate. She was full of smiles and giggles. We could tease her to no end. She had a great sense of play, whether it was singing silly songs, playing jokes on our eldest sister or parents, or acting out in imaginative play with "Alligators All Around" or some variation.

There was never any animosity with Suzie because with her there were no favourites. Whatever inappropriate words or actions we could get her to do on our behalf, we knew would not lead to a serious reprimand from our parents, because Suzie was so innocent and kind hearted. Our eldest sister even convinced her that she was born in a barn, a story she once relayed as fact to a doctor during an evaluation. She made us laugh and smile even when disgusting us in the back seat of the car as she would spit out the skin from apple slices.

Over the years, damage caused by illness and surgeries changed Suzie and our relationship. Gone was my bunkmate. Gone was my playmate. While I remember many stages along the way and many changes and challenges, one incident stands out in my memory as indicative of how far away I believed we were from the ideal of Suzie that I held on to.

After being away for a term at school on Vancouver Island, I came to visit Suzie at her new home in Benz Crescent in Murrayville. For me, accepting her living somewhere other than our home was a challenge in and of itself. I walked in and saw her sitting on the couch, eyes heavy and energy looking low. I leaned in to give her a hug (she had those great big bear hugs) and once I was embracing her, I enjoyed a moment of loving and being loved. But what I expected to be a kiss turned out to be a painful bite into my cheek. The bite hurt, but feeling like my kid sister was lost hurt more. The one who had always been so lovable and playful seemed gone.

Fortunately, that was only what it seemed at times. A few months ago, I got married and I had the joy of having Suzie, along with my eldest sister, standing with me. Suzie carried herself so proudly and participated in the activities, even as she murmured (not so quietly and in colourful phrases) that the reverend "better stop talking." She was exuberant on the dance floor, charmed my in-laws with her sweet, sociable ways, and like a true kid sister, finished her evening by taking the microphone and commanding everyone's attention by giving me a toast and a serenade of "Twinkle, Twinkle Little Star."

She's something special!

MEMORIES FROM SARA (ELDEST SISTER OF SAM AND SUZIE)

Suzie is intricately woven into all the good memories I have of childhood and life at home. Her laughter was contagious, and she always fell for my silly jokes and games. I remember putting all the chesterfield cushions on the ground and chasing Suzie around like we were bears. She was practically overcome with a mixture of laughter and fear—but never tired of the game.

I also recall Suzie, to our amazement, eating a pack of raw bacon, and a distressed Mom calling poison control.

She was generous to the point of giving all she had and could easily sing away hard times with "You Are My Sunshine." Suzie seemed invincible, despite her obvious challenges, and I still marvel how circumstances rarely dampen her resolve to accomplish what she sets out to do. God chose you for us and us for you, Suzie—you are loved!

CHAPTER 2

Medical History

In the first ten years of her life, Suzanna experienced a high number of seizures that significantly affected those close to her. They watched in desperation as the seizures affected her overall health and development. Disappointment followed by hopelessness filled the hearts of Suzanna's parents and sisters as one medication after another failed to bring the desired effect.

The doctors decided to reach for surgical measures in an attempt to control the seizures by removing damaged brain tissue. When Suzanna was only two and a half years old, she underwent her first brain surgery at the Montreal Neurological Institute and Hospital (1984). The following surgeries took place in 1985, 1989, and 1991. All of them would help to control seizure activity, but only for a brief period of time. They also resulted in the undesired side effects of a behavioural nature, as the frontal lobe was affected.

When all measures known to doctors at that time had been tried and had failed, the time came to try the most traumatic and life altering procedure—right hemispherectomy (1992).[4]

This was a life saving measure for her and a last resort. The doctors agreed that the sheer number of seizures she was having at that time would have literally "fried her brain" if nothing was done. Hundreds of seizures per day, uncontrolled and unresponsive to anticonvulsants,

4 For the benefit of the reader who is interested in the technicalities and details of what such surgery entails, a piece called "Functional Hemispherectomy" is included in Appendix A.

had contributed to the gradual decline of Suzanna's mental and psychological development. It was also believed that if a decrease in the number of seizures was achieved through this surgery, there could be a reduction in episodes of aggressive behaviour. The doctors were never quite sure if behavioural episodes were a result of seizure activity or of lost impulse control following previous brain surgeries or of both.

When Suzanna was ten years old, she had to travel to the Montreal Neurological Institute and Hospital (the only hospital in Canada that offered this kind of treatment at that time) for the procedure (her fifth brain surgery) that would leave her half paralyzed and with severe brain injury.

The doctors explained to the family in detail how this would affect Suzanna. Whatever she had learned so far in life would disappear from her memory and would have to be re-taught. She would lose abstract thinking ability, impulse control, and the use of the left side of her body. She would not be able to read, write, and solve math problems, sing, dream, or plan for the future. The hope, however, was that she would then be seizure free and would relearn how to live.

When Suzie returned from Montreal, she was sent to Sunny Hill Health Centre for Children in Vancouver for rehabilitation of the partial paralysis of her left side. A leg brace was prescribed to lessen knee hyper-extension while walking as well as for increased stability and support. A hand brace and physiotherapy were also prescribed.

The surgery did not bring the improvement that was hoped for. The seizures and extremely aggressive behaviour continued. When, due to behavioural challenges, the hospital could not keep her any longer, her parents were faced with a heart-wrenching decision. Instead of returning to her family home, Suzanna would move into Benz Crescent, a respite care home, where she would receive one-on-one support around the clock.

Suzanna mostly experienced nocturnal seizures. She had grand mal seizures two to three times a night, each lasting approximately two to three minutes. Protocol from Dr. Farrell had directed her caregivers to call 911 if a grand mal seizure continued for more than three minutes. There was always an overnight staff member with her. Suzie also had seizure rolls of shorter tonic-clonic[5] seizures, during which she remained conscious in between. The rolls would be when ongoing seizure activity took place, each seizure lasting about thirty seconds with seizures close together. Staff members had guidelines on these seizures as to when to administer Ativan, and authorization would be given by phone by Suzanna's parents, the supervisor, or senior staff. Ativan was given 2 mg at a time under the tongue, up to 8 mg in a twenty-four-hour period. If seizures did not stop, senior staff would call Children's Hospital and they would authorize another 2 mg. and usually, this would end the seizure roll. The tonic-clonic seizures were quite violent, and Suzie would have fallen off her bed if staff had not held her. She would also have a roll of seizures that would manifest themselves by a repeated severe turning of her head to the right. At one time she was in this mode for 104 seizures before they stopped. Again it was a judgement call as to when to administer Ativan. Even though the seizures did not resemble grand mal or tonic-clonic, it was nevertheless seizure activity.

Despite all the seizures, Suzie got up and went to school in Ladner, carrying on with her academics, work experiences, volunteer positions, and sports. It was always amazing to the staff that Suzie could have seizures so many times throughout the night, be given Ativan, and still be alert in the morning. Of course she was tired, but with encouragement she did carry on. It was important to those close to her that she not simply sleep her life away.

When Suzie was sixteen, Dr. Farrell ordered a brain scan with dye injected intravenously as she began a seizure. We did this at Children's Hospital, having to keep her amused and calm until a seizure occurred.

5 **Tonic–clonic seizures** (formerly known as grand mal seizures) are a type of generalized seizure that affects the entire brain. **Tonic–clonic seizures** are the **seizure** type most commonly associated with **epilepsy** and **seizures** in general, though it is a misconception that they are the only type.

When the seizure started, we could push the plunger to release the dye into her bloodstream. She would then be taken to a scanner where the staff could observe the dye going to the area of the seizure activity. It was going to the right hemisphere, which had supposedly been totally disconnected via hemispherectomy. With newer and better technology, it was discovered that there were spaces where the disconnection was not quite complete.

Suzie was scheduled to have her sixth neurosurgery done at Children's Hospital in Vancouver. There was now an excellent neurosurgeon on staff there. The old incision pathway into her brain was followed to complete the right hemispherectomy. During recovery time at the hospital, the incision opened up for some reason and had to be closed by the doctors.

The seizures continued, and Suzie was having so much difficulty that Dr. Farrell spoke to Bob and Cathy Bailey about the idea of trying a relatively new procedure known as Vagus Nerve Stimulation (VNS).[6] Similar to a pacemaker, this VNS device is used to treat epilepsy when medications are not successful. The stimulator device, about the size of a two-dollar coin, is implanted under the skin in the upper part of the chest. A wire under the skin connects the device to electrodes implanted in the brain or attached to nerves that go to the brain. The doctor programs the device to generate pulses of electricity at regular intervals. The hope is that these regular signals help prevent the electrical seizure outbursts in the brain. The individual is also given a large, hand-held magnet, about 1 x 2 x ½ inches in size, which when brought near the stimulator can generate an immediate current of electricity to stop a seizure in progress or reduce its severity.

Unfortunately, after many months of dedicated efforts, any positive results were negligible, and when the battery finally died out, the

---

6 **Vagus Nerve Stimulator (VNS):** VNS therapy is designed to prevent seizures by sending regular mild pulses of electrical energy to the brain via the vagus nerve. These pulses are supplied by a devise something like a pacemaker.

decision was made to stop the VNS treatment. To add a little humour to the Vagus Nerve Stimulator program effort, the magnet became Suzie's favourite throwing object to be hurled at staff whenever she was particularly upset about something. If she happened to connect with her target, it *definitely* left a bruise!

By this time, Suzie had been given every anti-seizure medication in every possible combination as well as investigational drugs. Some of them had undesired side effects (for example, excessive fatigue or increased aggressive behaviour), which outweighed the benefits. Some of the drugs simply did not work and were discontinued.

At one time Suzie was taken to Langley Memorial Hospital by ambulance due to prolonged seizures. The first time Suzie was taken there she was treated with the intravenous anti-convulsion medication Clobazam and sent home. The following night she was given an increased dose of the same medication. Suzie had been on Clobazam when she was younger, but there was no effect noted at that time. However, this time it seemed to work! Clobazam was reintroduced in an increased dose, and suddenly the seizures stopped. Something was working. Hormones? Past adolescence? Clobazam doing its job? Dr. Farrell opted not to change anything.

Suzie remained seizure-free for about thirteen years, approximately from age seventeen to thirty. By seizure-free, we mean no tonic-clonic or grand mal seizures. However, some form of seizure activity always continued. Suzie called it "shiveries," and the doctors were never quite sure if they were seizures or panic attacks. Even though they are not believed to be hurting her brain, they annoy and scare Suzie to no end, as she remains conscious throughout their duration.

Around the time Suzie turned thirty, "shiveries" started becoming more involved, sometimes turning into rolls of 100–200. We would also witness rolling of the eyes or a blank expression, but Suzie would remain conscious and talk to us. Feelings akin to panic attacks would frequently accompany these rolls. About the same time, Suzie started experiencing another form of seizures, which is now defined as focal seizures. They are nocturnal, like the tonic-clonic, and are characterized by stiffening of muscles and loud vocalization. We embarked on a journey to try to find treatment for them, trying both anti-anxiety and anti-seizure medications, but to no avail.

Thinking back over the years, we realize that as a child, Suzanna lived with seizures and accepted them because she did not even realize what it might be like not to have them. They were a sad part of her reality since before she could remember. She did not complain, and after a seizure she would pick up where she left off when it started. Having lived through all those seizure-free years and now having the seizures come back with a vengeance, she is able to tell us that they drive her crazy and scare her terribly.

The focal seizures come in rolls, and several trips to the hospital in the middle of the night have been made when Ativan failed. The doctors seem at a loss as to what more they can do for Suzie, so the staff team makes every effort to spare Suzie the stress of hospital visits and tries to keep her comfortable at home.

Reflecting back on the predicted consequences of the hemispherectomy, leave it up to Suzie to prove the doctors wrong! Some things did, of course, come true. The path is not easy, but she has far exceeded any expectations and is a living testimony of the brain's amazing ability to learn again. Remaining parts of the brain compensate for the function of the missing ones and allow Suzanna to beat the odds.

Was it worth it? Of course we will never know the "what ifs," but the surgery did decrease the seizures significantly, and the remaining seizures have been controlled by medicine successfully for many years following the surgery.

Suzanna did relearn many important life skills and is living a very full life. She needs one-to-one, around the clock support, but with assistance she is able to contribute to her community, to hold a number of volunteer and paid part-time jobs, and just "make everyone's day." She reads, writes, solves math problems with a calculator, sings to music, dances, and plans for the future.

Those of us who are close to her are proud of her achievements and celebrate her accomplishments.

Goldfish seizure

BY SYLVIA DOANE

When Suzie was about twelve years old she had a fish tank, which she loved and helped take care of. One day when cleaning the tank, one of the goldfish jumped out of the net and fell to the floor, landing on its head. The fish was stunned and not moving but came to life when returned to the tank. Thereafter, it would swim on its side. The fish seemed happy enough, but every now and then its body would shiver and shake. Then it would return to swimming on its side. We would laugh along with Suzie saying it had a seizure disorder too. Just like Suzie, the fish did not allow a few seizures to interfere with its life.

WEIGHT LOSS CHALLENGES

One of the undesired side effects of the various seizure medications Suzie was on was excessive weight gain.

Some of the medications increased her appetite tremendously, while others slowed down her metabolism.

On the other hand, her post-surgical left side paralysis prevented her from doing the more intense forms of physical exercise she was fond of as a young girl. As a result, Suzie's weight as a young adult reached a whopping 244 lbs.

The concern for Suzie's wellbeing and a fear of obesity-related diseases prompted her family and staff to brainstorm ways to help Suzie return to a healthy weight range. It was determined by her doctors that for her built, height, and bone structure Suzie should weigh no more than 180 lbs.

Initially, a point system was implemented. There was a menu plan prepared to help Suzie and staff create simple, healthy meals within her allowed point limit. Some staff found it easy to follow, while others struggled with following the guidelines.

When no significant weight loss was noted after several months, another weight loss system was implemented.

On this plan, Suzie made weekly trips to a local weight loss centre, where she was weighed, and could ring a loud bell to celebrate even the smallest weight loss. While there, Suzie purchased pre-made, pre-portioned meals for the coming week. She lost 23 lbs after several months on this plan, and then reached a plateau.

Other than the benefit of the weight loss, another significant advantage was that Suzie learned about portion sizes. She did very well with a predictable routine of meals served, and stopped seeking food outside of meal times. There was also several draw backs.

Due to the low nutritional value of the pre-packaged meals, Suzie's nails became brittle and her hair matted and began falling out. Her bank account was drained as the cost of food proved to be too much for her modest budget.

Around that time, Dr. Farrell suggested trying a version of Atkins diet to control Suzie's seizures. This diet was a modification of a keto-genic diet Suzie was on as a child when her seizures were out of control. Another brainstorming time of family and staff produced a low carb/ high fat meal plan. Suzie "graduated" from her previous weight loss program with a diploma, and switched to the new plan.

While there was no noted improvement in her seizures, Suzie quickly lost another 20 lbs and reached a plateau again. Her overall health, related to nutrition, improved significantly. Suzie now has healthy nails, and a head full of beautiful, thick and healthy hair.

She is incredibly proud of the weight she lost; something she does not hesitate to share with everyone she meets. She tells people that she lost weight by choosing healthier foods and by walking laps on the track. She is always applauded by her audience with high praise, which puts a big smile on her face.

Suzie is still looking for ways to lose the remaining 25 lbs. It is something that is very important to her and her parents. Preserving her mobility and decreasing stress on her already overworked joints (due to specific gait Suzie walks with) will surely only serve her well as she ages.

Staff and family continue to support Suzie in her goal, always brainstorming ways to adjust her meal plan and strategizing to optimize opportunities to exercise.

Many of us who struggle in this area admire Suzie's determination and self-control, and applaud her efforts.

JUNE 1992 THROUGH AUGUST 1997

Medical summary taken from Suzie's Individualized Service Plan, 1998

The first year at Benz Crescent was one of trial and error as staff, working five hour shifts initially to avoid burnout, endeavoured to learn everything they could about Suzanna's medical condition and her unpredictable behaviour. Suzanna was given the investigational drug, *Lamotrigine,* for seizure control in addition to her other seizure medications of *Tegretol, Dilantin,* and *Mogadon.* There was no change in the seizure activity, so *Lamotrigine* was discontinued. At this time, the decision was made by Dr. Farrell and the Baileys to remove *Dilantin* and *Mogadon* as well, considering the length of time Suzanna had been on these drugs.

Suzanna had previously taken horseback riding lessons, but was asked not to return to them as her behaviour was deemed too dangerous for that setting. She was often aggressive in the car, both to the car and to the driver, drastically reducing the number of staff who would take her into the community. Sudden aggression to strangers in the community as well as a refusal to leave facilities such as the library or a store proved difficult and challenging for staff. School was still only thirty minutes a day, with Suzanna's desk now placed in the cloakroom to minimize disturbance to other students. School entries became more difficult, and being with other students was a safety issue.

In October 1993, a team meeting was held at Shaughnessy Hospital, including interested persons from the medical field, social services, education, Langley Association for Community Living, and family members and friends. Questions were raised, brainstorming done. As a result, a psychiatrist from Children's Hospital was scheduled to see Suzanna. A letter read at this meeting from Dr. Farrell described Suzanna as "the most disabled child I have ever seen."

After assessing Suzanna, the psychiatrist tried separately and over a period of time three behaviour medications: *Ritalin, Prozac,* and *Clonidine*. Suzanna experienced rapid speech and movement, rage behaviour, and lethargy/drowsiness during this time as these medications were introduced and withdrawn. *Clonidine* was selected as the best possibility, and a period of experimental dosage regulation was begun.

In October 1993, Suzanna was suspended from school for aggression towards a teacher. Staff turnover at Benz was high as violent behaviour took its toll. It was critical to seek professional help from Gateway Behavioural Support Services and to apply for Suzanna's acceptance into the Gateway School Program.

Behavioural support was initiated in June 1994, and school admission was scheduled for September 1994. The new seizure medication, *Gabapentin,* was also introduced at this time.

A dedicated, professional, and cohesive staff team had been forming at Benz Crescent. Behaviour management programs were developed by Gateway School and implemented by Suzanna's staff, who became proficient in this area and continued to modify, improvise, and train to ensure success. The same program was to be followed at home and at school. Progress was slow and often discouraging;

however, specialized skills were developed by staff as they persisted and, working closely with the school, consistency and continuity were provided for Suzanna. During this period, Suzanna experienced many and varied ongoing behavioural and community challenges both at home and school.

During 1995/96, serious fatigue began to plague Suzanna. Seizure activity increased, behaviours persisted. Sleep patterns became periodically disordered. Repeated tests could produce no answers. It was medically decided that the only hope for decreased seizure activity was further neurosurgery. No medical answers were available for the fatigue. In 1994, Suzanna was referred to Dr. Hurwitz (psychiatrist/neurologist) and, after a long waiting period, admitted to UBC Hospital for seven weeks in January 1997 for observation and assessment.

She was scheduled for neurosurgery to complete the hemispherectomy in Vancouver at Children's Hospital in May 1997. Recovery was slow with repeated setbacks. Seizures persisted, fatigue and apparent depression controlled Suzanna's life. July was the low point. August realized dramatic improvements in the three areas of seizures, behaviour, and fatigue. She was able to begin the school year in September of 1997.

THE YOUNG SCOTTISH DOCTOR

Dr. Kevin Farrell, Suzie's Neurologist, has been walking alongside the Bailey family for nearly thirty years. He was the "young Scottish doctor," as Bob and Cathy Bailey fondly recall, to whom Suzie was first referred at fourteen months of age. It was at Children's Hospital in Vancouver that he continued to see Suzie throughout the years.

After Suzie became an adult, the going joke was that she could not be transferred to another neurologist, because her file was so big it filled an entire room in the hospital basement and was simply too large to be transferred. So Suzie continued to go to Children's Hospital until the day Dr. Farrell retired. She was over 30 years old at that time.

We, "The Book Team," had the pleasure of meeting with Dr. Farrell on April 4, 2014. He invited us to meet with him at Jericho Tennis Club, where he is a member. He treated us all to a lovely lunch and shared his memories of working with the Bailey family. We were all quite taken with this lovely doctor, who asked us to simply call him "Kevin."

During our meeting, Dr. Farrell talked about the impact Suzie had on him as a young physician:

> "I realized that helping the family cope is as important as the medical diagnosis. Most people go to the doctor for the cure, not for the diagnosis, and I couldn't offer them the cure. Suzie was fourteen months when I saw her. I had to help the family grieve "the loss of the perfect child." As a doctor, ninety per cent of the time you diagnose a problem but can't do much about it. You support the family and help them reset expectations. When I met Suzie, I knew nothing. When we train doctors, we make sure they are competent and that they don't kill patients. The most important part is our experience, but experience comes from the mistakes we make. It's a lifelong learning exercise".

When Suzie was born, personal computers had just been released. There was no e-mail and no Internet. There was no MRI in Vancouver, only in the specialized hospital in Montreal. Technological advances have made a huge difference in terms of curing/treating epilepsy since then.

When you meet a specialist who says, "I don't know much about this, but," he is the one you should go to. It's the "know it all's" that should be avoided.

The key to the success of Children's Hospital's Epilepsy Clinic is the nurses and nutritionists, not the doctors and specialists. Every day you have different issues to deal with. The critical role of the nurses is their availability, their listening ear.

A supportive team is the key. It takes a village to raise a child. When it comes to Suzie's staff team, it has also been exceptional how she has been looked after. Another young man I support has a similar set of challenges to Suzie's, but his success has not been as evident because his team was never as consistent and dedicated as Suzie's. Staff consistency and the ability to "pass the torch" played a huge role in Suzie's success.

Suzie forces you to reflect on what's important in life, what your values are. This is a gift we get from her. Suzie comes in, and you suddenly realize, "I don't have any problems!" One in twenty children with epilepsy come to a point when epileptic processes slow down or stop the development of the brain. This is a devastating fact, but you cannot give up. People can develop all the time, but you have to remain open to development. Suzie taught us that and proved the science wrong.

As her parents often mention, Suzie has been very blessed over the years to be looked after by such remarkable doctors and nurses who saw her for the person she was, entangled in the tragic circumstances of her medical needs, and not just a medical case. They all, and Dr. Farrell in particular, went above and beyond the call of duty to support the family on the path they had to walk.

Sebastian

BY SYLVIA DOANE

Suzie would sometimes babysit her parents' dog, Sebastian, when they went away for a few days. When Suzie would start to seizure and yell in her bed, Sebastian would come running down the hallway like a bullet. He would jump on top of Suzie, barking his head off right in her face, and all we could do was sit on the bed and wait for the seizure to end. Suzie would, of course, thank Sebastian for taking care of her.

TIME AT THE UNIVERSITY OF BRITISH COLUMBIA PSYCHIATRIC UNIT

By Sylvia Doane

Jane Huff and I supported Suzie at UBC Psychiatric Unit, where Suzie was admitted in 1997 for a psychiatric evaluation.

Dr. Hurwitz was a Psychiatrist and Neurologist, and my understanding of his introduction into Suzie's life had to do with behaviours and how to deal with them. Dr. Farrell (Neurologist) suggested Dr. Hurwitz, who was the head of Neurology and Psychiatry unit at UBC Hospital. Dr. Farrell said he liked the fact that Dr. Hurwitz "thought outside the box."

Suzie was admitted into UBC's psychiatric unit for a seven week assessment. When Dr. Hurwitz came into Suzie's room for the very first time, she picked up her hair brush, and quickly threw it at him, and made a direct hit. Dr. Hurwitz just said, "Good morning Suzie and Jane!"

They were very long days for both Suzie and staff. The shifts were sixteen hours, and the two- hour drive to and from Langley made it into an eighteen-hour day for both Jane and me. We worked one day on and one day off alternately. It took tremendous energy, creativity, skill, and patience to successfully fill those long hours. Dr. Hurwitz visited Suzie nearly every day. With the exception of taking Suzie's temperature and blood pressure, we saw little of the nursing staff.

When I was at the hospital with Suzie in the morning, I tried to establish a classroom-like schedule. For example: Wake up—Breakfast—Shower—Get dressed. This could take up to a few hours. Then: Snack time—Reading—Math—Colouring—Singing—Walking (which Suzie did not care for) around the UBC grounds. One of the activities Suzie really enjoyed was setting up a spare room for music, dancing, and singing. Then: Lunch—Out for pop and a walk. This was rest time for the patients. Suzie, of course, did not go along with this "rest stuff." She had not rested since she was a baby. During this time, we sometimes had a picnic lunch away from UBC.

The Psychiatric Unit was not developed nor intended for a young patient like Suzie. I also found the setting to be dark and drab. Patients walking the halls would sometimes stop and talk to Suzie and sometimes not. At times this would become a problem. When hospital staff tried to intervene, they were met with some of Suzie's most colourful language. Soon enough, most hospital staff learned to leave us to ourselves. They did not have the knowledge or skill needed to safely support Suzie and her behaviours. Suzie got along with most of the cleaning staff, who would take the time to listen and talk to her. Suzie made her own bed and tidied her room. She did not understand why people did not want to converse all the time. There was one exception. A lovely older lady sat on a bench outside the hospital almost every day, always with a wonderful warm smile for Suzie. They would sit together for fifteen to twenty minutes while Suzie told her all about the news from yesterday. Then we would move on to get our pop, but not before Suzie and her lady friend said a long goodbye, which included prayer and songs. In the many times this took place, the lady did not say a word. She just smiled and nodded intensely. Oh, the joy of these special interactions for both of us!

The one and only time I left Suzie alone was when two of the nurses convinced me to take a break and have lunch in the cafeteria next door. When I returned half an hour later, Suzie was not in her room. When I checked with the nursing station, I was told Suzie had been assisted into the timeout room by two of their staff. I looked into the small window and there was Suzie, sitting on a mattress on the floor, bent over with her head in her hand, not making a sound. My heart ached. They told me Suzie had become aggressive, swearing and pushing one of them. I asked to have the door unlocked, and they said it was unwise, as I would get hurt. I asked again to open the door and lock it behind me! I walked over to Suzie, sat down beside her, and put my arms around her. She then burst into tears and I did too. We sat like that for a while and when Suzie was ready, we went back to her room.

It was the saddest thing! I was told later that one or both of the nurses had to take some stress time off.

At the end of the seven week assessment period, it was so nice to see how much Dr. Hurwitz cared for Suzie. On the day she was being discharged, Dr. Hurwitz dropped by to say goodbye to Suzie and her staff. As we met with him, we were expecting some final directions regarding Suzie's care. Instead, he surprised us by saying what a wonderful job we had done setting up such a successful program. He talked about how much he admired our positive approach, offering Suzie choices and always speaking in a soft and respectful manner. He spoke to us about how consistent we were as a staff team, recognizing how helpful this was in allowing Suzie to make good decisions and in turn feel good about herself. Dr. Hurwitz mentioned that he was so impressed with our approach to assisting Suzie that he was considering setting up a program in the hospital that incorporated a similar set of working principles.

His parting suggestion to us was to keep up the good work, offering no additional advice. We left feeling good and very proud knowing that through trial and error, we were on the right track.

Hot and salty!

BY JANE HUFF

When Sylvia and Jane were doing the alternating sixteen hours shifts with Suzie at UBC Hospital, they managed to communicate about the important things they needed to share about each day. They did not really feel they needed to add in anything personal.

Sometime later, after that seven week time, Jane (somewhat shamefacedly) happened to mention that every night after her long shift ended and she left the UBC parking lot about 11:30 p.m., she drove directly to the McDonald's located near the exit from the campus. There she ordered large McDonald's French Fries—fresh, hot, and salty. Often there was no opportunity to eat throughout the sixteen hours at the hospital, so by that time those fries (HOT and SALTY) took on a gourmet delight. It fortified the soul for the trip through the nighttime road construction on the freeway back to Langley.

Sylvia began to laugh and said she had done exactly the same thing. In fact, for the last couple of hours of each shift, that was all she could think about! Oh, those hot, salty French Fries! This identical sneaky behaviour had never been mentioned until that time later, but both minds had certainly continued to run on the same track as they wound down from the long days. Neither had resulting heart attacks.

CHAPTER 3
Education

THE EARLY SCHOOL YEARS
MY TIME WITH SUZIE
Memories by Marge Blake

I was the teacher in a Special Education classroom at Douglas Park Elementary School in Langley. There were ten children (aged six to nine years) with various special needs in my class. I also had two and sometimes three full-time teaching assistants. The administration, staff, and students were accepting and supportive of our program and students. We were included in all school activities such as assemblies, sports days, holiday celebrations, and concerts.

In the 1980s, Langley was an ideal place and time for families with children and for children with special needs. There was an active Langley Association for the Mentally Handicapped, (now LACL) a special needs pre-school, parent support, and parent advocate groups in the community. The Langley School District had established special needs classrooms with strong support from district personnel and school administrators. The wonderful programs, facilities, and support already in place were the work of earlier parent and community advocates. For Suzanna and her classmates, the program at Douglas Park Elementary was the first step of many that led to where and who they are today. In Suzanna's own words, "Marg Blake, you are not going to believe it!"

I first heard about Suzanna Bailey in June of 1987. I attended a meeting with Cathy Bailey (Suzie's mother) and the Langley School District Head of Special Education to discuss plans for Suzie starting elementary school in the fall. Suzanna was in the hospital at that time, and I was unable to meet her. I learned that she was five years old, had frequent severe seizures, had undergone several surgeries, and was on numerous medications. She would be six in August of that year and hopefully would be able to attend school.

That September, Suzanna joined our class. She came in bright-eyed, chatty, and enthusiastic. This was Suzie when she was not having or recovering from a seizure. That first year, when Suzanna was well enough to be at school, she had seizures about 85–90 per cent of the time. As a classroom teacher and with the support of three teaching assistants, it was possible to provide a safe, appropriate, and happy environment for learning. Activities and programs could proceed even when one or two of the staff was needed for the spontaneous health (i.e. Suzanna's seizures) and behavioural incidents that occurred.

Suzie was eager to participate in all lessons and activities. The school day was busy and structured. The focus was on the students participating in and learning academics, self-care, life skills, social skills, and community skills to the best of their ability. Suzanna learned in spite of her seizures and the busy surroundings. Often after a grand mal seizure, she would be flat out, barely awake on her mat, and still raise her head in an attempt to participate in lessons.

We frequently had district support staff (speech and language specialists, physiotherapists, audiologists, and psychologists) or specialists from Children's Hospital and Sunny Hill observing the classroom and assisting us in programming for the students. One of Suzanna's early assessments suggested that she learn to type, since printing or writing might be something she would be unable to do. However, if her classmates were printing, Suzie was printing! She was determined to be a part of all that was going on in class.

During the years that Suzanna was my student, she had three major surgeries. Two of these took place at the Montreal Neurological Institute and Hospital. I was able to go with Cathy Bailey for two of these events. After each surgery, I was able to spend a couple of days at the hospital visiting Suzie. Within minutes of arriving, I realized she was recovering and would soon be back at school. Even major surgery could not hold her back.

Suzanna's behaviour constantly changed as she experienced numerous changes in medication and physical growth. The surgeries also had an impact on the potential for her behaviour to become worse. At times she was docile, compliant, and exhausted from the effect of seizures and medications. Other times she would be boisterous, active, and challenging. She could express herself using the most acceptable, kind, and loving vocabulary. She was also capable of using the most colourful language, including cursing and swearing, which she told her teachers she learned from her dad and her sisters. Meanwhile at home, she told her parents she had learned the swearing from her teachers. Regardless of the challenges, Suzanna appeared to enjoy school and she continued to learn. A day in the classroom with her was never dull.

Operation Track Shoes is a track meet that takes place annually at the University of Victoria. The track meet is open to children and adults with special needs from around the province. This amazing event is organized and carried out by volunteers. It takes place from Friday to Sunday on the second weekend in June.

The staff and students of our special education class at Douglas Park attended Operation Track Shoes for a number of years. Each student was paired with a volunteer and was billeted in the dorms at the university. The volunteers were with the students for the weekend, accompanying them to meals, track events, and social activities. Teachers and teaching assistants accompanied the students to the events and were always available to assist and support them. The parents of the students (especially when a child was attending for the first time) often came to join in the fun and to cheer the children during events.

Suzanna, along with her classmates, was able to participate in this weekend event. She was fortunate during her first year to have a registered nurse as her volunteer. Suzanna took part in track events, swimming, and social activities such as magic shows, dancing, and a banquet. The result of one surgery had left Suzanna with less strength and mobility in one arm. The supervisors were concerned that she would not be safe competing in the swimming event. A decision was made that she could swim, but that she had to have a strong swimmer swim beside her. Suzanna got into the water, flipped onto her back, and proceeded to swim independently the two required lengths of the pool. She graciously accepted her ribbon amidst cheering from teary-eyed spectators.

On one occasion I met with Suzie's parents. Just one look at them told me that they had had no sleep. With Suzie and two other children to care for, they were exhausted. My heart went out to them, and since my children were away at school/university and my husband worked weekends, I offered to take Suzie home with me for the weekend. My teaching assistants offered to be on call should I need them. I could not sleep for the first two nights. I came to realize just how many seizures she had during each night.

Having her overnight or for a weekend was a learning experience for me. It was a different experience than teaching her in a classroom for five hours with support staff. Initially, I was concerned about her having seizures, especially at night when I was on my own with her. I had not realized how much time and attention she needed. Suzanna was at her best when she was active and busy.

Later, on occasional Fridays, Suzanna would come to my home for a night or two. It was just Suzanna and me—no support staff and no planned/structured time. The hours seemed much longer. Other than meals and bedtime, we did not follow a routine. Suzanna liked to be busy and did not want to do things alone. She was always eager to participate in an activity if we did it together. On rainy/indoor days and for down time before her bedtime we did quiet activities such as reading, music, colouring, cutting, puzzles, and activity books (matching, dot-to-dot, etc.). Suzie liked helping with household chores, setting the table and baking, and especially stirring. She also enjoyed folding laundry and getting mail.

During fine weather we were outdoors. We took lots of walks to parks and playgrounds and around the neighbourhood. She especially liked White Rock Beach and Crescent Beach. Being on the beach was always special. When it was warm we would walk barefoot in the sand and water. We flew a kite if it was windy. Suzie liked finding shells and stones and watching birds. Sometimes we went to a restaurant. Most often we had a packed lunch and beverages with us and ate at a picnic table. I have fond memories of Suzanna's footprints in the sand as she wandered and explored the beach.

This time in our relationship caused me to see, despite her seizures and other limitations, her zest for life. I feel fortunate that I had the opportunity, ability, and energy to spend this time with her. I certainly developed a greater understanding of the challenges faced by parents and families of a child with special needs.

Suzie was in my classroom from age six to ten. After her last major surgery in Montreal, she was ten years old. She stayed at Sunny Hill Health Centre for rehabilitation of her left side. She was so aggressive that there was no possibility she could come back to the classroom or to her family home.

In September of 1991, I took a teaching position in Delta School District at Gateway Provincial Resource Program. I taught for a year in the residence and then moved into a classroom at Delta Senior Secondary, an outreach extension of the Gateway program. Suzanna was a student in my classroom once again. Jane Huff worked with Suzanna, assisting her in the classroom as well as transporting her to and from school each day.

PLANNING MEETING

After Suzie had been living at Benz Crescent and supported by LAH staff for eighteen months, a meeting was called. This meeting was intended to provide information for future planning in all aspects of her life and developmental needs. A variety of people were involved in the meeting. This included her parents, medical team, educators, and caretakers.

A manager for LAH (now LACL) opened the meeting by stating that Suzanna had experienced both challenges and successes over the past year and that this meeting would consist of sharing information only on the challenges that Suzanna faced.

Suzanna's parents, Bob and Cathy, summarized their situation by observing that Suzanna had unique problems as well as incredible talent and ability. They felt that most people could not see the person in Suzanna, only her behaviours and problems. Bob and Cathy wanted answers as to how Suzanna's strengths could best be developed.

What follows is a summary of the reports discussed at the meeting.

- The "Oh Suzanna, a Nursing Challenge" account was handed out and presented by the nurses at Sunny Hill. A video showed Suzanna's left side weakness and poor concentration after surgery.
- A Pediatric Neuropsychologist, having completed assessments on Suzanna in 1983, 1985, 1987, and then again in 1993, stated that her greatest problems were related to the neurological damage that occurred during surgery as well as her inability to do motor planning. He observed that even though she was twelve years old, she functioned most often in the five-to-six-year-old range. A comparison of assessments showed very slow progress and that there appeared to be a plateau of growth. He noted that Suzanna could be quite pleasant and cooperative once she had settled in to any given event. Her rote memory skills, which assisted her in some of her learning, were very good. He also noted that she required a highly structured environment.

- Dr. Farrell, Neurologist at BC Children's Hospital, wrote a letter. In it he stated: "Seizures are not well controlled. Efforts are being made to control the seizures with a combination of medications that will produce the least number of side effects." He acknowledged the extreme severity of Suzanna's handicaps but could not determine if these disabilities were due to surgery or seizures. He had a psychiatrist make a home visit to discuss Suzanna with her parents and her caregivers.
- An Occupational Therapist had supported both staff and Suzanna in the use of activities in daily living as therapy for Suzanna's left side weakness. Suzanna experienced a lot of difficulty in dealing with visitors to her home environment. The therapist saw an increase in Suzanna's disinhibited behaviour, which decreased the amount of time spent on physiotherapy exercises. Staff were using counting and reading to decrease these behaviours. With these techniques, Suzanna was able to better verbalize her problems, particularly those connected with wearing an arm/hand splint at night and foot/leg brace throughout the day. Although Suzanna did not like wearing these supports, she responded well when Jane Huff, her staff and teacher, helped her deal with the exercises and goals. The therapist saw a need and benefit for therapy to continue on a one-to-one basis. It had to take place in Suzanna's home, and consistency would be vital to her success.
- A Counsellor had completed an assessment on Suzanna's suitability for counselling around anger management and emotional development. She identified a number of issues that needed immediate attention:
    - The educational system could be expected to provide or determine Suzanna's suitability for counselling.
    - Suzanna's disinhibited behaviours made her very vulnerable in her community.
    - Suzanna needed training in sexuality and coercion.
    - Suzanna lacked skills for appropriate peer relationships.

- ▹ Suzanna's support staff experienced tremendous demands from Suzanna and needed a system of support for them to continue to work effectively.
- ▹ The school system was not receiving the professional support services needed to effectively plan or evaluate their programming.

While the school system did not feel equipped to meet Suzanna's educational or social needs, it was the one source of appropriate peer groups available to Suzanna.

- ▸ An Assistant Principal at the Langley School District reported that Suzanna had started attending Murrayville Elementary in October of 1992. Although the students were well prepared for her arrival, Suzie was only able to attend for half an hour each day, and although her classmates were eager to work with her, she was only able to participate in a few extracurricular activities. That year, Suzanna had been unable to meet the expectations for classroom behaviour. She attended school for half an hour three times a week with staff and was seated in the cloakroom with her one-on-one worker. Suzanna's isolation and the use of some anger management strategies allowed her to continue to attend school. Her classmates grew increasingly afraid of her, and Suzanna's behaviour presented a safety risk to the other students in the school. It was recommended that planning address the following issues:
    - ▹ development of a positive peer culture
    - ▹ development of an extensive and ongoing behaviour management program
    - ▹ provision of a specialized education service that promoted both academic and social growth
    - ▹ provision of an age appropriate environment and service with attention to her developmental needs.

The 1993 school year was seen to be failing to attain the goals as presented by Suzie's parents, teachers, and staff.

After further discussion about the future of her education, it was suggested that Gateway Services for Autistic Children be explored. This agency operated a specialized school and residence for children and young adults who were unable to be successful in a typical school setting due in part to behavioural challenges.

Gateway Services were located in Ladner, and if Suzie was accepted, it was decided that Jane Huff would drive her as well as accompany her to class each day that Suzie was able to attend.

There was a long waiting list, but the assessment requested of Dr. Farrell could suffice to speed up the process. Gateway also required that families have a Microboard[7] to ensure funding. The Ministry of Social Services and Housing had knowledge of Gateway's services but had not previously dealt with Microboards. Another hurdle to overcome was that staff in the school needed to have a teaching certificate. It was a "light in the tunnel" when Jane Huff, who had accompanied Suzie to Murrayville Elementary School, announced that she held the appropriate qualification to attend school with Suzie each day.

At this point in the meeting, alternative residential options and recommendations were discussed. It was identified by the MSSH that future funding at this level was not assured. Other options for shared accommodations and staffing had to be explored. There was also concern that this model did not offer Suzanna the opportunity to be with her peers. It was agreed that Suzanna needed to live in a home that had at least two staff on duty during waking hours. She would need one staff member at any time she was in the community and at times when she was agitated or upset. Suzanna was as big as or bigger than all of her current support staff and could be harmful when she lost self-control. The second staff person needed to be present for safety purposes and team support. This home would need sufficient funding

7 A Microboard™ is a small (micro) group of committed family and friends (a minimum of five people) who join together with the individual to create a non-profit society (board).

to allow for weekly staff meetings as a means of addressing problems promptly and guarding against staff burnout.

In the face of such extensive service needs, it was important to remember that Suzanna was a child who, like any other child, required a caring and nurturing environment. One goal would be for Suzanna to learn to share common living areas, but she would also need private space in which to complete activities requiring concentration and attention. When the MSSH suggested a specialized foster home, the Baileys instantly replied that Suzanna would not need foster parents. She had parents who loved her but who needed support at this time in order to continue to advocate for her care.

At the end of this period of time, the Executive Director of LAH was successful in ensuring that MSSH recognize Suzanna's unique residential needs: the importance of her staying in Langley in an environment of full support *and* that funding continue to support her at Benz Crescent during these pivotal years of personal growth and development.

It was also determined that Gateway Provincial Resource Program would be worth trying in order to see if this would be a better fit than her neighbourhood public school. The only drawback was that the school was in Delta and a significant distance away from Benz Crescent.

Suzanna's education was so important to the staff at Benz that arrangements were made to ensure that she and her staff (Jane Huff) would be welcomed and supported. *This is an example of how a collective group of caring people helped Suzie develop into who she is today.*

GATEWAY SCHOOL

By Jane Huff

When it had been determined that the only school Suzie might fit into was Gateway in Ladner, we began that journey. Suzie and I travelled the seventy-five km round trip each day to school. Classrooms were small and we focused on the Gateway curriculum with its components of appropriate behaviour in the community and some academics such as money recognition.

Gateway had two rules: keep your hands and feet to yourself and use nice language. That covered it all. Suzie could respond to a reward system, a sticker for each increment of time following the rules. When she reached a certain number of stickers, I would take her to a wonderful little deli named Elizabeth's in Ladner.

Staff supported each other at Gateway. Suzie could have lunch in the Gateway dining room with a certain teacher she liked very much if that morning she had not thrown things at her overnight staff. Suzie became much better about not throwing things after she knew she could have this special lunch time.

In the Gateway program, Suzie went on group excursions such as bowling, swimming, and tobogganing in the North Shore Mountains. She and I would fly down that hill and trudge back up pulling our toboggan. We would warm up with hot chocolate and go down again. It was a fantastic experience. She did all of this despite her disabled left side and while wearing her leg brace.

Suzie had her first work experiences with Gateway. Once a week she vacuumed the warming hut at the Reifel Bird Sanctuary. It was also a favourite place to go to hike, feed chickadees from our hands, and just enjoy the beauty of wild wetlands. Suzie was also able to have a work experience at Kentucky Fried Chicken in Tsawwassen. She dished up the coleslaw and macaroni salad into containers and put on the lids. She learned the rules about working in a food service: Wash your hands. If you touched your face or your hair with your hands, wash them again. She cleaned the kitchen at JR Direct, a mailing service in Ladner. She cleaned a large glass partition and the countertops, put things away, and did whatever needed to be done. Suzie also had a work experience at the Boys and Girls Club in Ladner. She washed the windows, upstairs and downstairs. It was a lovely facility. Occasionally there were some seniors' events there. If that was the case, she could help set tables for a luncheon when she was finished with the windows.

When Suzie went from the Gateway school building to the Provincial Resource Program (PRP) classroom at Delta Secondary School, the reward system was the same for all students. A chart had the day divided into time increments. Every increment that had a check mark for keeping hands and feet to self and using nice language was worth 10 cents. At the end of the day, the total amount earned was charted, and Friday was "pay day." Students were then given the choice of spending the money in Ladner in the afternoon or taking it home. Gateway residents could spend it with a staff member during the weekend. I believe the top amount possible for a chart full of check marks was $2.00. It was a great incentive for most students. Parents were billed for the incentive money since there was no money budgeted for this.

Cues were ongoing for Suzie. She was always attracted to mothers walking with their babies or small children. She would want to sing all the verses of "Hush, Little Baby" when she met them on the sidewalk. Some mothers were quite enchanted with Suzie singing all the verses of this song. Others were a little nervous and wanted to continue on their way. We began to work on how to greet strangers in the community. I would ask her to say "Hi," and to keep on walking. We worked on this until she was successful. She would need to be cued in advance (i.e. "Do you see that lady coming this way? Can you say 'Hi'?"). Sometimes it was just too hard and she had to wave her hands and say, "You just won't believe it," one of her favourite phrases. We would keep practicing and achieving more and more success. The way she would look at me with that proud "I did it" look and a happy smile when she was successful was always a joyful moment.

Then there was the time when we saw a man in a phone booth. I asked Suzie if she could walk right by and just say "Hi." She said she could. We proceeded nicely and the man jumped right out of that phone booth and initiated a conversation: "Well, good morning ladies, and how are you today?" Of course Suzie had to tell him he "just wouldn't believe it."

During this time, Suzie did not have many behavioural incidents at school. She was very focused on her work. She loved doing school work and there were great resources both at Gateway and the PRP classroom.

Suzie did have a hard time with transitions. As much as she liked her classroom, getting into the classroom without an uproar could be difficult. At the high school, we had an outside door in our classroom as well as the door that went into the main hall. Getting into the school, down the hall, and into the classroom via the inside door was hard for her. She would walk faster and faster, beginning to "rev up." It was not just us. We were entering with hordes of teenagers going into the school. I did not want the other students in the school to be afraid of her, so I asked if we could move Suzie's desk beside the outside door so she could go and sit down straight away when coming in that way.

To avoid the loud and high energy entry into the classroom, I had to get her to focus totally on something. I used Suzie's natural kind and helpful nature to assist me with putting a Band-Aid on my "hurt" finger every morning. I would scratch my finger across the stucco on the school building as we walked around to the side door so that it was a little red. I had a Band-Aid in my pocket, and every morning I asked her just at the door if she could please help me because I had hurt my finger on the rough wall. Every morning she said she could and she did. Right into the classroom with Band-Aid in hand, right to her desk at the ready, and she was in. I had my finger tended to. In a way, this was one of my favourite strategies. It worked consistently, and we were able to fade out the Band-Aid strategy. After a while she could just go in and sit down.

Suzie was attracted to the "Goth" kids, the ones with the black capes, dreadlocks, and pierced noses. She found them "pretty gorgeous" and told them so with a big smile. They loved her and were the nicest kids. When we were on our way into Ladner to go to the post office with the school mail every day, kids out playing soccer and football would wave and greet her. I never experienced any of the high school students being mean or negative towards her. Teenagers can be so great!

At school, whether doing academic work or at a work experience, it was imperative that Suzie have time to finish what she was doing. If that was in place she would most likely have a successful experience. If the task was not made to fit into the time period, things would most likely all break loose. I would help her so that she could finish within the time slot and everything went smoothly.

OPERATION TRACK SHOES

By Jane Huff

Operation Track Shoes is an annual track meet held at the University of Victoria in Victoria for special needs children, teens and adults. My first experience with Track Shoes was with Gateway and then with the PRP classroom. We went as a school group with T-shirts, a schedule of participating events, and high anticipation. When we marched on to the field to the music of bagpipes along with hundreds of other athletes, it was a sudden teary-eyed moment for me.

Suzie was a teenager our first year there and was able to attend the teen dance and banquet. Every year she participated in high jump (not too high), long jump, shot put, and 50-metre run, hurdles (stopping at each one and carefully stepping over), and swimming. We had to hustle to get from one field event to the swimming pool. Suzie had her own counsellor (with me two steps behind at all times). She had her usual nighttime seizures but was up and ready to go the next day.

There were dorm accommodations the first few years and then we were given our own apartment with our own bathroom. This was in student housing, just a short walk from the main campus. This worked very well and helped keep Suzie on an even keel.

I do not recall any major behavioural issues other than Suzie giving her very lovely counsellor a good swift and sudden kick for no apparent reason. According to Suzie, she just deserved it. Counsellors are well trained at Track Shoes, and sudden behaviour from time to time is not unusual.

Operation Track Shoes involves an amazing number of volunteers. The extensive training they receive enables them too safely and with grace, enthusiasm, and class support all these special people with such varying, amazing athletic abilities. It showcases sheer determination and guts and has many elements of pure fun.

I will mention Suzie in the swimming race the last year I took her. She was wearing a new electric-blue swimsuit. I carefully apprised the supporting people in the pool wearing wetsuits that Suzie would jump in the deep end at the starting gun and turn over on her back to do the backstroke with her strong right arm. I told them that she might need slight support under her left armpit and to keep her in her own lane, but that she did not need any other assistance. The gun went off, Suzie jumped in, then proceeded to baffle me by turning onto her tummy and doing a very strong right-armed crawl with much thrashing, splashing, and one-sided swimming. It was one of the most amazing things I have ever seen. Not only did she do it, but she came in first! She climbed out, put her electric blue ribbon over her head, and brought the house down. Success! Again!

DISNEYLAND
By Jane Huff

When Suzie attended school in the Provincial Resource Program classroom (PRP) at Delta Secondary, several years of fundraising resulted in the entire class being able to go to Disneyland. The daily entry pass and money for lunch each day were the only individual expenses. Ladner fishermen had donated smoked salmon to be sold, and firefighters had donated money. Hamburger and hot dog sales were held at local supermarkets, and packaged chicken was sold. These were just a few of the fundraising events. A local radio station rented vans for the entire group upon arrival in Los Angeles. This allowed the group to have transportation to Anaheim and for the duration of their stay.

It was quite an amazing feat to have raised enough money to cover flight, motel accommodations, and insurance as well as breakfast and dinner for everyone each day. The documents at that time were simple: parental permission, ID, and birth certificate. No passports were required. There were no incidents on the airplane with our group of students with autism or with Suzie.

When we arrived at LAX we went outside to a kind of cement enclosure to await the arrival of the vans. Suzie looked around with awe and wanted to know if this was Disneyland. Our motel was a nice one fairly close to Knott's Berry Farm. Suzie and I had our own room and there was a large facility that seated everyone for breakfast and dinner. We had purchased breakfast cereal, bread for toast, and juices for breakfast. For dinner we ordered in different things such as pizza, chicken, vegetable and fruit plates, salads, soups, hot dogs, and spaghetti. This was cost-saving. Everyone liked the food, and this prevented the students from always having burgers and fries. They could have whatever they wanted for lunch at Disneyland. The motel provided us with some toasters for breakfast toast and some cooking facilities for the hot dogs and spaghetti. This worked and everyone ate quite well, considering that this was not home.

Upon arrival inside the Disneyland gates, we proceeded first to the courthouse on Main Street, USA, for our passes to go directly on the rides. We did not wait in line and that was a great thing for special needs individuals! Suzie and I went on our own way every day, meeting the group at the vans at the appointed time in order to return to the motel.

One evening we had tickets for a Vancouver Canucks-Anaheim Ducks hockey game at The Pond. Suzie happened to have her Canucks hat with her and wore that to the game. I was not sure what she would think of attending a hockey game, so I took my bag with colouring, calculator and math sheets. She astounded me by paying attention to the game the entire time. We did not even open the bag.

Suzie loved the rides at Disneyland. I think we did them all, including the haunted mansion and some high and fast ones. It was nice to be there for a few days and do everything as well as her favourite rides several times. Other than a few minor outbursts when we were all in the dining hall at the motel, Suzie was great the entire time. We were grateful to all those who worked so hard to make the dream a reality.

Surrey Arts Centre

BY SYLVIA DOANE

Suzie and I went to the Surrey Arts Centre to see a Christmas play. At intermission, Suzie needed to use the washroom. The cubicle area was too small for the two of us, and Suzie suggested that she could manage by herself. The bathroom was crowded, so I waited a little ways away, close to the sinks. In a few minutes I heard, in a loud and clear voice, "Sylvia, I had a large b.m." This led to giggles all around. A few moments later came Suzie's voice again, "Sylvia, don't forget to write it down!"

10 THE LANGLEY TIMES Friday, February 8, 2002

LETTERS

Susie's satisfaction an inspiration

Linda Drake,

Credit to Black Press and Langley Times

CHAPTER 4

Residential History

TRANSITION INTO RESIDENTIAL SERVICES INTERVIEW WITH DAN COLLINS,

Executive Director of Langley Association for Community Living (LACL)
By The Book Team
(Added comments from Jane Huff and BettyAnne Batt)

Suzie was ten years old when Dan first met her. Dan always came to staff meetings and supported the staff. Bob and Cathy also felt very supported. Their family was always made to feel welcome and their assistance was appreciated. They continued to monitor and take charge of all medical decisions and appointments, just as if Suzie was still living at home with them.

DAN: The family asked our agency to support them in their need for a service for their daughter and to ensure it was one that would keep her close to them in Langley. Benz Crescent, an individualized residential service model, came as a result of this request.

This service was before its time and, as a result, was thought to be too expensive. However, the family was desperate for assistance, and we were able to acquire initial funding for a year. We knew that Suzie needed to live alone and she needed an individualized program with highly skilled, carefully selected staff if she was to have any life at all.

There was a constant battle with funding bodies because they had not previously supported this model, and did not fully understand the need for it. We had to appease them from time to time and give it a shot, such as the group living arrangement. Not every organization would have taken this battle on and seen value in fighting for what's right.

Every year the ministry would say, "You have to come up with a different model. This model of support is too costly." This request was made annually by the funders. I just kept saying to them, "Yes, yes ... I will explore other models of service" and did what I needed to do to ensure Suzie's needs remained paramount. But they kept coming back. The first year, and most years thereafter, we as an agency expended more money than we received, but we never stopped providing Suzie and the Bailey family with an exceptional service.

JANE: Her mother said her dreams for Suzie were that she'd one day sit at a table and that she would have a friend. One day Suzie was in a restaurant sitting with friends, and her dad cried. He said he never dreamed that Suzie would live to that point.

DAN: I maintain that nothing has changed. We have to maintain the priority of keeping specialized staff focused on helping individuals live their lives to the full potential. Not a lot of people get the art of our work. I had to keep alive the hope that those folks (specialized staff) would eventually show up and they did.

JANE: Suzie was just a little kid, and I wanted her to have joy like other kids do. I remember taking her tobogganing, and as she was flying down the hill, I saw the look of pure joy and delight in her face. Sylvia Doane taught me not to give up if it didn't work the first time. We tried and tried again.

BETTYANNE: In those early days the most important thing was to remember that Suzie was a child. Staff was always available to support

Suzie after hours. They didn't mind taking phone calls from Suzie. We had to take a risk, had to take a chance; otherwise Suzie would not have a life. On Suzie's seventeenth birthday, Suzie came in and didn't see anybody in the crowd but her dad. Everything froze. Suzie's emotions took over until she "got it all out," meaning shouting everything negative that was on her mind. Then, and only then, she was ready to start over with a more appropriate greeting and conversation.

DAN: The community can now read Suzie's emotions. They know her now. It used to be that "the sea parted" when Suzie came in. Everyone was afraid of her. These days, people in the community are drawn to Suzie and see her gifts, talents, and enormous heart. I don't know that we can put a price on the depth of our connection with Suzie.

LIFE AT BENZ

By BettyAnne Batt

I met Suzie for the first time in her classroom at Douglas Park School. Although I was there to meet another child, I also wanted to meet with Suzie and her teacher, Ms. Blake. I knew that Suzie's parents were in need of some meaningful respite. It was near lunchtime and Suzie invited me to share her lunch. Suzie was six years old at that time; however, medical intervention put the plan of respite on hold. Suzie was to have another significant surgical procedure to try to calm her seizure activity.

Over the next couple of years, Suzie would be in and out of hospital, trying new medications and forms of treatment. In between exhausting seizures, she would be experiencing challenging behaviours. After her trip to Montreal for her most intrusive surgery, Suzie spent many months in Sunny Hill Health Centre for Children. It was in planning for her discharge that her medical team at BC Children's Hospital and Sunny Hill strongly recommended that she be placed in a home that would continue with her rehabilitation and offer her parents and siblings some respite.

Suzie had lost most of her impulse control. Her behaviours had become overbearing and constant. She was impulsive and unpredictable. Many feared that she would hurt them as her rants and anger overcame her sweetness. Her sisters remember having to have a special lock on the door of their home because she would run out of the house. Anything that could be thrown had to be put away, out of reach. Her personal safety was at risk at all times. She had lost the use of her left arm and wore a brace on her leg. She required physical rehabilitation and therapy.

The executive director of Langley Association for the Handicapped (LAH), was summoned to a Ministry of Social Services and Housing (MSSH) meeting to hear firsthand the intensity of the situation and of the need for family support that would allow Suzie to come back to Langley as an outpatient of BC Children's Hospital.

The MSSH Manager for Langley and the Baileys' Social Worker for Children with Special Needs had requested a meeting in the conference room at the Langley MSSH office. Here a proposal, which would assist in bringing Suzie home from Sunny Hill, was being presented for funding approval. This would offer a rented home in Suzie's home neighbourhood, fully staffed around the clock and supervised by LAH, as a means of providing rehabilitation for Suzie and respite for her family. This was to be a six-month contract.

At that meeting, I said: "Suzie Bailey is truly a very complex individual. Those who have come to know her recognize that the potential for personal growth exists where there is a strategic plan that is creative, responsive to the support needs of the Baileys as a family, is innovative, effective and holistic. Should this residential plan be successful and set a precedent for other children with complex needs throughout British Columbia, then let the process begin with those who share this vision for the future of Suzanna Bailey."

Needless to say, we were all greatly relieved to hear when immediate planning could begin. Very significant to this endeavour were the leadership skills and sensitivities that the management of LAH contributed.

The proposal allowed for a temporary home to be established for Suzie. Funding came from the Family Support and Respite Care Program of LAH. We immediately advertised for seven staff members to facilitate five-hour shifts for each twenty-four hours. We received twenty-two applications and interviewed them all. Over the next few months, we advertised and re-hired several times. We were looking for a kaleidoscope of personalities to contribute to the support of Suzie and her family and to build into the house at Benz Crescent an extension of "HOME."

One of the key staff members we hired was Jane Huff, a very kind, gentle, but strong teacher and parent. That she had a heart for Suzie was immediately evident. Jane soon became the only constant in Suzie's life at Benz. She became a mentor and role model for all other staff as well as a strong support for the Baileys. Jane took over the educational tasks of teaching Suzie and of setting up protocols for other staff to be successful.

Shift changes were tremendously stressful for everyone. New relief staff and those staff who did not understand Suzie's need to be cued and prompted continually were unsuccessful and often left in fear.

It took a while, but some trends were showing the staff and family a path whereby situations could be made more positive for all. Soon another significant staff person, Sylvia Doane, came to Benz. She was fun-loving, respectful, and offered Suzie outlets for developing her artistic creativity with decorations and celebrations that were simply amazing. Sylvia's compassionate approach soon became another role model that enlightened the new staff and supported the existing staff team.

Prayer Time

BY SYLVIA DOANE

When Suzie arrived at Benz Crescent she was already in the habit of saying her prayers at bedtime. This is important to Suzie, and even now she continues this nightly routine. Suzie learned the Lord's Prayer and a few other prayers along the way. She would always begin by saying "Gone Jean Lord (meaning God, Jesus Lord), please keep the angels around my Mom, my Dad, and Sara and Sam, and all my family and friends." As time went on, the list of acquaintances started to get longer and longer, and everyone she knew had to be mentioned in her prayers from time to time. We would talk about getting a handle on all the names and start from the beginning … this worked for a while, and then the names were added one by one. It continued this way for years and years. There were times when Suzie would not agree with this strategy, get mad, and simply state that "I HAVE to say all the names." While other nights she would cheerfully agree with the suggestion that if the prayer was too long, she would never get any sleep and be very tired in the morning. The easy compromise was that Suzie would mention the name after the lights were out and all was forgotten. Thank the Lord! Suzie always ended her prayers asking the Lord to "Take away these blasted seizures in the Lord's name. Amen!"

BY SYLVIA DOANE

When Suzie first arrived from Sunny Hill she slept in a twin-size bed and wore a restraint. This was the same restraint that was worn during her stay at the hospital. It was believed this was necessary to keep her in bed rather than have her running around the house throughout the night. If left without the restraints, Suzie was a threat to the support staff's well-being and/or would do damage to the residence. She would also be up throughout the night. When she would finally go back to bed, the possibility existed that she would sleep in until noon or later.

The staff team was uncomfortable with the use of restraints and set a goal to support Suzie without them. The details of this goal were only finalized after much discussion over many meetings.

To enlist Suzie's understanding and acceptance of this idea, we set about talking to her and emphasized all the positive results. When asked if she would like to sleep without restraints she said, "Of course I would!"

We proceeded to explain the following aspects of our plan to help her achieve this goal.

- Suzie would move into the master bedroom.
- The room would be painted a colour Suzie would choose.
- New curtains (again, Suzie's choice) would be purchased.
- Suzie would sleep on a new queen-size mattress with new bedclothes and nice furniture.

Everyone was excited about the goal except (understandably) the overnight staff. They would be left with the responsibility of being directly involved in trying to achieve this goal. To assist the staff person on duty, it was agreed that the rest of the staff team would take turns sleeping overnight in the home since back-up help might be required.

For the first few nights, Suzie got up continually and came charging and screaming down the hallway towards the night staff. In an attempt to discourage these outbursts, it was decided to block the hallway with a large armchair. This had minimal success. Suzie would get angrier and begin to throw things.

So it was back to the drawing board! We asked Suzie if she was unhappy with these changes and if so, would she like to go back to her former room. She was most upset with this idea and told us so. We came up with some additional strategies.

- We would leave the hall light on throughout the night.
- Suzie could call us at any time and we would go to her in her room.
- If Suzie would like, she could read to us from her bed.
- Suzie would be supported to say her prayers for as long as she wanted.
- We would have a night light on and play some soothing music.

Again, it took a few nights of praising any and all positive efforts Suzie made in following through to achieve the goal. The end result of this was that Suzie fell asleep without a lot of fuss.

Suzie was so pleased with herself and truly enjoyed her success as well as all the positive support she was receiving. She loved the idea that she was praised for being a grown-up young lady and understanding that is how adults behave.

It took about a month and then there was consistent SUCCESS! We were all happy for Suzie. We loathed the idea of the nighttime restraint from the very beginning. It took some time, but after much trial and error, and only after many hours of planning and strategizing, the goal was achieved.

This is an example of the successful conclusion to a goal that provided Suzie with a positive change and improved self-confidence.

Honeybunch the cat

BY SYLVIA DOANE

As an older teenager, Suzie lived at Benz Crescent where the bathroom was located right across from her bedroom. When Suzie was finished bathing, she would wrap a towel around herself and head into her bedroom. We would then watch Honeybunch, the cat, jump up on the toilet (seat up) and put her front paws right down to the water and have her own bath, lasting for five to ten minutes. Yes, she was a super clean cat!

Tom is very proud thinking they made a path in the snow just for him.

"It's time for my bath now," said Honeybunch.

JODY MEADOWS, SUPPORT STAFF

I began supporting Suzie on Saturday evenings, Sunday mornings, and Monday evenings. The first thing that I wanted to establish was a relationship with her so that we could feel safe and confident with each other. My shifts provided me with this opportunity.

On Saturday evenings after dinner we went to Dairy Queen, with the goal in mind to make a good choice. It took some time, but we discussed what a good choice was by looking at the calories, etc. We had established that Suzie needed to lose some weight. I did too, so it was a great idea to reinforce a good choice. Suzie handled this so well, even to the point that she would explain it all to the counter help every single time. After our treat we would walk/drive to the video store to pick up a movie. Getting the exercise and helping me lose weight was always a factor in Suzie's decisions. I loved my Saturday evenings with Suzie.

Suzie would make me porridge and had it down pat how to prepare and serve it. After getting cleaned up we would take a drive, and sometimes her sisters would come for lunch. She would also paint my fingernails, and to this day she is the only person I have allowed to do this. She liked and still likes colouring in colouring books. I tried to hold off on this activity as long as possible so that it would not consume our time together. Sundays, as I recall, were filled with talks about my family and hers. It was a wonderful opportunity for her to see who I was as a person and for me to get to know her.

Suzie and I began to get to know each other well through rhythmic gymnastics. It presented lots of triggers and opportunities for behaviours that challenged Suzie and me as well as the gymnastic participants. We talked about her entering the gym. (She still makes the best entrances.) We discussed that she would tell me when and if she had had enough. Sometimes it went well and other times, not at all. The important thing is that we were going through it together and we were building our respect for one another. Suzie was seeing that I could help her when she needed me the most. I remember good times, laughs, and giggles. Suzie made friends and they remain as such.

As the years went by, Suzie and I understood that we were together for the long haul. She became a very important part of my family, watching my children grow up. We supported each other through volunteering at St. Catherine's School library, where we put books away, at Otter Co-op shredding paper, swimming, Kentucky Fried Chicken salad preparations, Meals on Wheels, at St. Nicolas Church folding bulletins, grocery shopping, doctor and dental appointments, and her last surgery.

I believe the common thread for the success we have had together is that I presented myself as not a perfect person. I was learning and still am. I apologized when I could have been better.

Ditch Diving
BY JANE HUFF

One day, Suzie and I stopped at my house on our way to Benz Crescent. It was wet and slippery because of wet leaves in the driveway outside the gate. Suzie got out of the car on her (passenger) side and somehow slid gracefully into the ditch in front of our house. The ditch had about six inches of running water in it and is probably three feet deep. Suzie could not climb out. I jumped into the ditch and tried to push her from behind. No go.

I knew there was no one at my home to ask for help. I did not wish to leave Suzie alone in the ditch, because surely a police car would drive by and find Suzie alone, stuck in the ditch.

In a few minutes a car came by. We flagged it down and a very nice man came to our rescue. He grasped Suzie under the arms and pulled. I was in the ditch pushing up from behind. We got her out! Suzie thought this was the best joke ever and laughed and laughed, even though she was soaked and cold.

She was fantastic in the midst of a strange adventure. When we returned to Benz Crescent, I removed my soaked shoes. I seemed to be tracking black dye from my shoes onto my socks and feet and then onto the Benz carpet. Just one of those days!

Special Olympics

BY SYLVIA DOANE

As a young teenager, Suzie did not have friends outside of school, and her staff team wanted to find ways that would encourage friendships for her.

I brought up at a staff meeting that it was time that Suzie join a Special O activity, and I suggested bowling as a start. There was silence in the room. It was apparent that staff were fearful of not being able to manage this activity, and so I ended up taking her on Friday night for the first time. She was to be part of a girls' team.

I was terrified, as I had no idea what would take place. My biggest fear was that she would throw the ball at someone. The first thing that triggered Suzie was when the ball went down the gutter. She would get mad and swear loudly. The other girls and I decided to cheer every time the ball went down the gutter because nobody else could do it as well as she could. Another trigger was when one of the girls would say "It's your turn Suzie" She would tell them to shut up, it is none of your business!"

We survived that night but week after week there would be other triggers, some that you could not foresee or plan for. The girls on her team were most upset at her swearing. I took the time to talk to the girls and explain that Suzie could not help herself, and that I needed their help. As the weeks and months went by, we encouraged Suzie, and we were successful in getting her to say "Oh darn it" when the ball went into the gutter. To this day, she will repeat that phrase each time her ball goes to the gutter.

Suzie and I started to bring a treat for all the girls each Friday evening, and it did not take long before they accepted

Suzie as she was. Week after week, a wonderful friendship developed between Suzie and her team.

When Suzie went up to bowl, I had to stand behind her in case she lost her balance or needed me to redirect her. I was never more than a foot away from her as at any time she could swing out or hurt somebody. In addition to the many incidents that happened at bowling, Suzie would often have seizures, and her team would be so helpful to me.

Bowling has always been a high threat activity for staff and for Suzie. To this day, it still causes a lot of stress for all, but the risks are outweighed by the pleasure she gets from bowling.

Suzie and I were also involved in Special O bocce, rhythmic gymnastics, baseball and disc throwing teams.

Special Olympics has always played a role in Suzie's life as she enjoys parties and events put on for the athletes, and the many friends she has made over the years.

RESIDENTIAL PLANNING AFTER GRADUATION.

As Suzie was fast approaching adulthood (aged 19+), the funding model would change from the Ministry of Children's services to adult funded services. It was time for the family and professionals to think about where Suzie could live successfully as a young adult as her time at Benz Crescent was coming to an end.

As is the case for most young adults, a group home environment was the leading option presented. The adult ministries were not in favour of funding an individualized residential model of service (like Benz Crescent) at this time.

A PATH was a planning tool that involved all those who were able to share their dreams and goals for Suzie's future as a young adult. Where she would live, work and play, as well as where her medical, social and behavioural needs would be met would be included in this document. It was a process that took several meetings to complete. This tool would guide many decisions that were very important to her parents. Having friends to share your day with was a primary goal.

After much planning, Suzie moved into Tall Timbers residence, a staffed group home operated by LACL. She would be living with two other housemates. Initially, everything seemed to be going well, as the three individuals were getting to know each other and settle into their new home. Families were also transitioning along with their young adult children in support of this home and model of care.

The staffing model allowed each individual to have their own support staff, each specifically trained to meet the individual's unique challenges and support requirements. However, as time moved on, it became increasingly evident that attempting to support these three very different individuals and their unique needs/programs under one roof had many unanticipated limitations. Everyone involved did their very best to make the home a viable residence for all three individuals, but over time it became more and more apparent that the "meshing" of the individuals and their support staff was failing.

After many positive efforts to make the residential model work at Tall Timbers, it was finally determined that the issues and challenges would not be satisfactorily resolved, and arrangements were made to move these three young adults into smaller home environments.

At this time Suzie moved into her current townhome, purchased for her by her parents, with ongoing support and around the clock staffing provided by LACL. In the small townhouse complex called Woodbridge, she would get to know her neighbours and be able to continue living there throughout her adult years. For over a decade now this has been a wonderful place where friends and family gather for meals and visits. Suzie is so proud of her home, which she shares with her cat, Lucy.

Suzie's 17th Birthday

BY SYLVIA DOANE

On Suzie's seventeenth birthday, she asked if she could go to the Red Robin restaurant for dinner. It was quite the gala event with about twenty-five family, friends, and staff joining her to celebrate the occasion. Suzie was in her glory with lots of smiles and giggles as she visited with everyone. When Suzie entered, she did not see anybody in the crowd except her Dad and her emotions and excitement took over, as usually happens when she sees him, and she shouted, "Dad, get your ass over here!" as she charged towards him, arm waving and using some of her most colourful language. This ended, as it always does, with kisses and a big hug. When dinner was over and Suzie had opened the last of her gifts, Suzie's dad stood up and thanked everyone for coming. With a shaking voice, Bob talked about the joy and wonder he felt as he watched his daughter celebrating her birthday, laughing and having such a wonderful time. In reality, with the many seizures and surgeries and the limited use of her left side throughout her life, he had been unsure Suzie would reach her seventeenth year. "I believe it to be truly a miracle." With that he looked over at his daughter with tears in his eyes and said, "Happy Birthday, Suzie!" Looking back at him Suzie said, "Thank you, Dad."

"I love doing my school work!"

"And then came the day I graduated from Delta Secondary School!"

"I can make this jump!" Long jump at Operation Track Shoes

50m run at Operation Track Shoes

"I love decorating my house for the holidays!"

"Grouchy me. I don't do mornings at camp!"

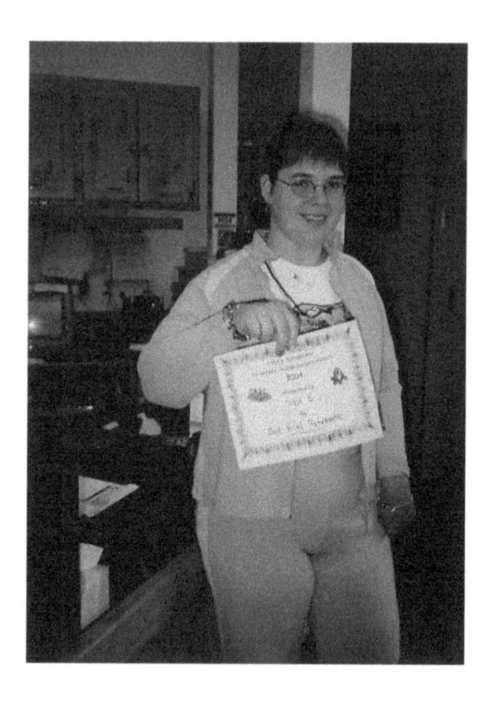

"Happy me. I love singing at camp!"

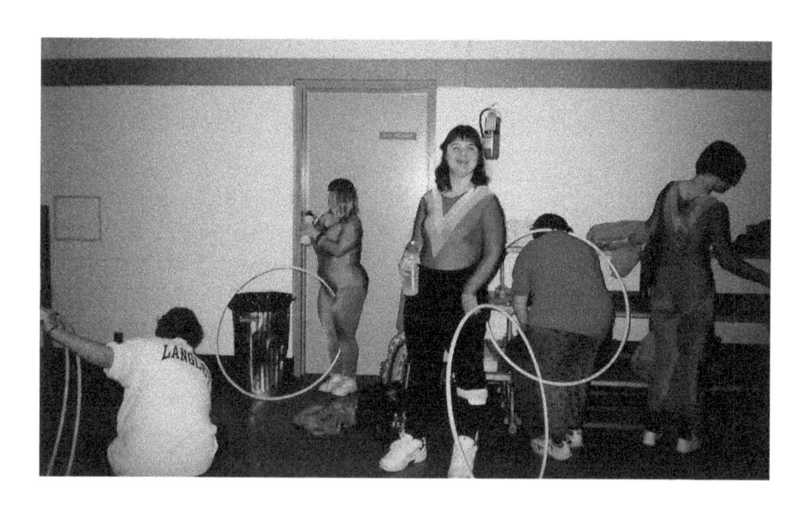

Rhythmic gymnastics.

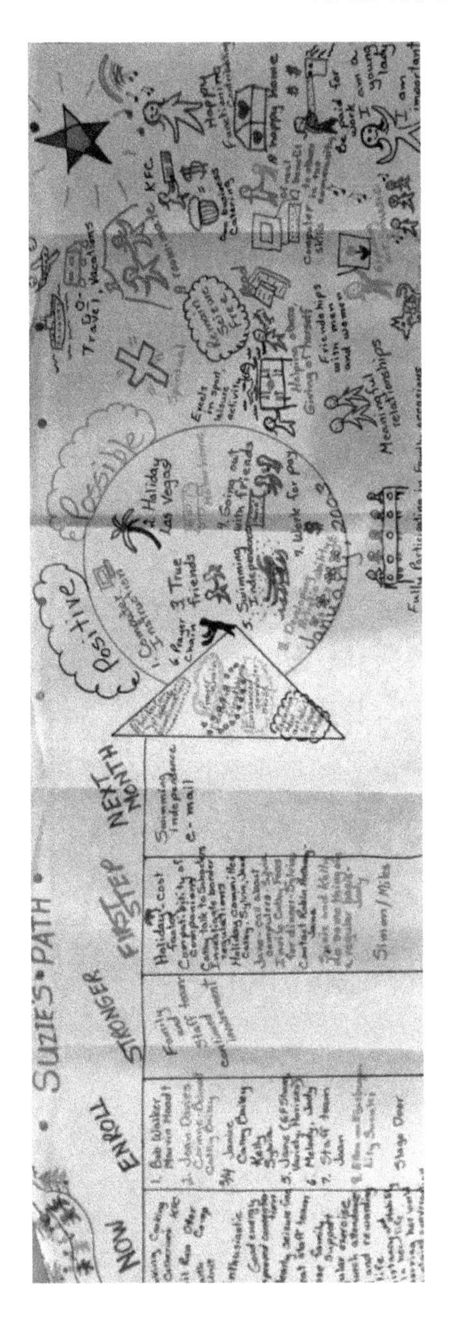
SUZIE'S PATH
NOW
ENROLL
STRONGER
FIRST STEP
NEXT MONTH
Positive
Possible
Holiday Las Vegas
3 True friends
Travel, Vacations
KFC
Meaningful relationships

Suzie's Path

CHAPTER 5

Challenging Behaviours and Strategies to Address Them

BY THE BOOK TEAM

As our writing was taking us toward the final few chapters of the book, we took some time to review how we described our early years with Suzie. On reflection, we quickly realized that we were not being totally honest in how we were describing Suzie's challenging behaviours. In truth, it was horrendously difficult in those first few years for everyone. There never seemed to be a day that went by without serious incidents or aggressive behaviour. This left many staff members questioning whether or not they could return the next day and start all over again, and many did not return. There were many times when two staff members working side by side were needed to ensure both Suzie and staff were able to get safely through the day. What kept bringing most of us back were those precious moments when you could see and recognize the other side of Suzie—the kind and loving child she tried so hard to be.

As a staff team, we quickly learned that we had to remove almost everything from around the house that could be accessed by Suzie. Household decorations and knick-knacks were put away. In the kitchen area, cupboards, drawers and pantry were emptied of anything that could be dangerous. Even during mealtimes, you could never turn your back to Suzie in fear of cups, plates, or utensils becoming flying objects heading your way. Her good arm was so strong that even essentials such as fire extinguishers had to be bolted to the wall. She also took to kicking walls when frustrated, and succeeded more than once in breaking through the walls, especially in the kitchen. During this time, there was a great turnover of staff, as many found it too difficult to cope with the stresses of the work.

We also took a cautious approach to being close to Suzie, as she might lovingly embrace you or, alternatively, take the opportunity to cause you harm. When Suzie was upset for whatever reason, she wouldn't just come your way swinging her arm, but would also try and kick you. If she was able to get within reach, she would try very hard to spit at you or pinch, bite, or grab your hair. If she managed to take hold of your hair, it was often pulled out by the roots. Her strength made it difficult to break away from the struggle, but the staff were always cognizant of not harming Suzie in doing so, as difficult as that might be. The altercations would leave you in tears, your heart pounding and feeling so very sad for both yourself and Suzie.

Suzie loved to be outdoors. This presented different, but equally challenging, behaviours requiring close monitoring and precautions. Part of the problem was the Benz Crescent home's landscaping. It didn't take long for Suzie to start picking up rocks and throwing them at staff, neighbours, and cars. For some reason, Suzie thought this was funny and delighted in the reactions achieved. She also had a habit of throwing rocks at staff through the patio door and cleverly loading her pockets with rocks to later walk into the kitchen and throw them there. Subsequently, it didn't take long for staff to park their vehicles out of Suzie's throwing range. The home's glass windows also felt the brunt of Suzie's rock throwing fun. They were replaced many times, with one permanently replaced with unbreakable Plexiglas.

Although the yard was fenced and despite careful supervision, Suzie still found the means to sneak away. When successful, her first effort was to run to the neighbour's house and barge in, surprising them as she went through the house screaming and kicking walls.

They were wonderfully patient and understanding of these occasional interruptions!

Our staff meetings were held weekly. These meetings always included the executive director, who was so supportive and involved in seeking out plans and options to counter Suzie's challenging behaviours. Thinking back, we all knew we were not going to give up. That was never going to be an option! Our mantra was "try and try again." We never left that planning meeting without new strategies in place, along with a positive attitude to keep us moving forward in Suzie's best interests. We were all committed to trying the new strategies consistently because we knew that was our best chance for success.

Ensuring that Suzie was as successful as possible at home, at school and in the community was a process. Behavioural strategies were learned and developed over time and by trial and error.

The principles mentioned below have been successful most of the time.

- When Suzie is out of control, those around her have to be at their best. This was repeated at every staff meeting and when introducing new staff to Benz.
- Stay calm, even when one's heart is racing and pounding.
- The authoritative way is easier, but it does not work.
- The soft method of teaching is very difficult but is more successful.
- Remember that this is a project of patience.
- Ask Suzie if she is ready to do something and then say, "Whenever you are ready." That empowers her to move forward and more often than not, she will be ready to tackle whatever is ahead of her.
- Remember that new situations are stressors for Suzie. The dances at Patricia Hall were an example of this. The first time she went, she had to go in and out six times before she felt she was ready to meet people.
- Remember that she has a hard time differentiating between excitement and aggression. In a new situation she may be aggressive, or she may need some time to do what she needs to do and then she

will respond appropriately. When she gets excited she may have to say one of her sayings, such as: "Those darn dogs barking are making me crazy," and then a positive conversation may follow.

- Cue, cue, cue.
- Be proactive, be proactive, and be proactive.
- Try again and again.
- Laugh and rejoice over success.

Some people have had a difficult time understanding Suzie. One staff member said, "I don't understand why she is like that." Suzie's aggressiveness and impulsivity follow her wherever she goes. If only people knew the extent and severity of the brain damage she suffered as a result of multiple neurosurgeries!

Once people get to know her, they realize that they cannot change her. They can begin to understand what she has to deal with, and that she lives with her inability to stop herself every day. It takes a lot of energy for Suzie to be able to deal with the emotions that come with her lack of impulse control.

Behaviours like hugging and then biting, pushing, and spitting require an understanding of how her brain reacts to emotions. Family members have always received the worst aggression; since they are the people she loves the most.

In the following pages, we've attempted a detailed description of behaviours Suzanna displayed and some of the strategies we developed while supporting her.

One of the purposes we had in mind in writing this book was for it to be a helpful tool to the families and caregivers of children/individuals with needs similar to Suzanna's. Out of our love and respect for Suzie, our hearts wanted to present you with a rosy picture of everything that's lovely, sweet, and good about her, but we realize that if we did only that, some of you may think: "What's the big deal? You guys had a cake walk compared with what I'm going through!" So, reluctantly, we made a decision that we would be frank and honest about the challenges and difficulties we faced along the way. The truth is, in those early days the behaviours were so bad that Suzie never went out, except for her doctor's appointments. Staff members continually brainstormed on when and how to introduce Suzie into the community. One of her first outings had been to the "Flower Garden," which was a lovely name for the cemetery. There were rarely any people there, and little that could be damaged, so it seemed a logical place to start. In those days, Suzie's parents' dream was that Suzie would one day be able to sit down to a dinner in a restaurant with family and friends, and one staff member hoped to be able to take her shopping.

Part of the frontal lobe of Suzanna's brain was removed during one of her surgeries. Consequently, she lost her behavioural "brakes." Literally, this means that the built-in ability to recognize and apply a socially appropriate response to any given circumstance is gone. This ability is replaced by anxiety, fear, difficulty transitioning, and impulses, which result in so-called "behaviours." Whatever crosses her mind, she acts on it.

In order for Suzanna to succeed, she has to be taught replacement behaviour continuously. She needs to understand the consequences of her own actions, both positive and negative. Once Suzanna gains that understanding and is able to sort out in her mind the order of her world, she still faces difficulty understanding variations to standards (i.e. how things are supposed to be and how to react when they are not). She still has to make a choice how to respond every single time. She still faces her underlying anxiety with every transition, be it a transition from one activity to another, from one support staff to another during shift change, or from waking to sleeping. Every situation or transition requires Suzanna to make a conscious choice of how she is going to react, what words and actions she is going to choose. She needs the constant support of family members or staff that must always be a step ahead of her.

Something Suzanna loved to do since she was a little girl was throw things. It seems that she threw items for several different reasons:

Sometimes it was to express her anger, sometimes when items were in her way or in a place where (in her opinion) they should not be and sometimes out of joy or purely for fun. Suzie threw everything from rocks in her yard, to ornaments in her house, to kitchen utensils and silverware. She got as creative with finding things to throw as it gets: fire extinguisher, thermostat just pulled off the wall, and even a sink. Where she found the strength to pull the sink off the wall with only one arm is still a mystery. She threw things at windows and shattered them, at people, at cars belonging to staff, and into the river.

It was important to keep in mind that Suzie had no impulse control, nothing to stop her on her own. She needed our help, our cues and hints. It always had to make sense to her and be consistent between staff and family. At first, in order to get a positive end result, we had to reward her with something she wanted.

One time, she threw the car registration papers out the car window on Hwy 10 while returning home from school. Jane Huff, who was with Suzie, had to stop the car, lock the doors, and retrieve papers from the ditch.

Jane came up with a strategy, which in this case worked on the first try. She told Suzie that the glove compartment containing the registration papers had to be locked for one week, and after that they would try again to leave it unlocked. Suzie never threw the papers out again.

The same strategy was used when Suzie threw cutlery or kitchen utensils. However, it took many tries before Suzie stopped doing this. This problem occurred more frequently with new staff.

Suzie used to throw rocks at people's cars and house windows.

As an alternative to rock throwing at property, Suzie was told that if she wanted to throw rocks she could throw them over the back fence into the vacant field behind the Benz property. If we visited a park, for example Williams Park, with a small river and no one nearby, Suzie was allowed to throw rocks into the water as much as she wanted. This seemed to work better than saying just no rock throwing at all.

Another thing Suzie loved immensely was cutting everything with scissors. Clothing, bedding, curtains—everything was in danger! As with throwing, simply saying "no" did not work.

The best strategy was to equip Suzie with scissors and ask her to trim the bushes surrounding the house. She could then cut to her heart's content.

From the very beginning, there always seemed to be a "weak link" among the staff team whom Suzie would identify and target. This took the form of going after this staff person by throwing objects like clothes hangers, hitting, swearing, and so on. It was not enough to try reasoning with Suzie and asking her to be nicer to the targeted person. If Suzie managed to have a good shift with this staff person, she would be promised positive incentives. These included lunch with a favourite staff or a community outing from a variety of activities she enjoyed. As a team, we always worked hard to create new and exciting choices as positive incentives. Although to a lesser extent, this problem continues to this day.

In Suzie's younger days, some of her daily behaviour included spitting, pinching, hair pulling, and kicking. Once Suzie was in an escalated state; she had no control over what she did and resorted to everything she could to hurt the person with whom she was upset. There was no single strategy to deal with those behaviours, but the violent episodes would definitely decrease in frequency, intensity, and duration as Suzie established a relationship with the person, getting to know and trust her better. As Suzie grew up, she was taught replacement skills and learned to use her words rather than her hands and feet. The episodes have become less frequent and some of the behaviours have stopped altogether. However, to this day, a lot depends on who is supporting Suzie before the "trigger person" comes on the scene and how they are able to prepare Suzie for the transition.

One behaviour that was particularly difficult to deal with while out in the community was bolting. For one reason or another, there were times when Suzie would suddenly feel compelled to bolt, for example, behind counters in stores. This was particularly frightening to clerks and was always red-alert time.

Proactive strategies included walking on the side of Suzie that presented the bolting opportunity.

Suzie would also bolt from her own house and run into the neighbours' homes to kick their walls. This happened particularly with relief staff who may not have been as attentive or attuned to what she might be doing.

If Suzie went towards the door, it was important to go with her, asking if she would like to go for a walk or perhaps do a little yard work, such as get those bushes trimmed!

Suzie had a work experience at Reifel Bird Sanctuary, vacuuming the warming hut. She would frequently try to bolt behind the counter in the gift shop.

The only response that worked was to block her with one's own body to prevent her from crashing behind the counter.

At the bird sanctuary, she would often throw herself on the floor, kicking and screaming. This may not have been such a problem if it wasn't for a fragile stained-glass hummingbird area. We did not want a hefty bill for broken stained glass!

We had to let it wear out, blocking the kicks towards glass, and then ask her if she was ready to go to the car. The Reifel manager would often help walk Suzie out to the car. After an incident like this, work experience was cancelled for a week and then resumed. Eventually the incidents ceased.

Suzie was still in a bolting mode during a work experience at Krause Berry Farms. She suddenly bolted through a door into the room with stacked crates of blueberries.

Jane had to block her from kicking over the blueberry crates. Suzie's kicking and screaming on the floor just had to subside on its own. Jane waited it out and then they left in a nice way.

One strategy frequently used in dealing with challenging behaviours is *redirection. Phone calls have always worked well for Suzie. If she was upset with the person supporting her, a phone call to a friend, family member, or another staff person helps her to gain control of her emotions and deal with the issue at hand.* Sylvia and Wayne Doane and Jane and Dick Huff have been faithful friends to Suzie over the years. They have taken calls from her whenever needed, regardless of the fact that any such call may take from thirty to sixty minutes. It certainly helped prevent many critical incidents and serious injuries to staff. It has helped Suzie to work through her emotions and to come up with better solutions to problems even to this day.

One of the characteristics very specific to Suzie is her unique use of spoken language. She loves to put a LOT of expression into the words she says, be it positive or negative. It is as if she wants to shock you. She will use those very words: "You will be so shocked," or "You just won't believe it," along with what she wants to say, even if it is something as simple as going to bed in the evening.

Suzie may start off with sharing something negative like: "You won't believe what this stupid, farthead, asshole house has been doing to me! It has been trying to make me wobbly and trying to make me fall down," or "You will be so shocked! Those stupid pervert shivery seizures were bugging me a hundred times!"

It is then up to the person speaking with Suzie to empathize with her and steer her back into a positive mode.

The conversation might go like this:

"Listen to me, you big fat asshole! This stupid, fart head house has been playing tricks on me and trying to make me fall down!"

"Oh, I'm sorry to hear that, Suzie. What did you do to deal with that?"

"I tried to hold on to the wall so I would not fall down, but the house was still doing this to me."

"So what did you say to yourself, Suzie?"

"I said, 'Do my best and forget about the rest.'"

"That's so smart of you, Suzie, to say those things to help yourself. Was there a staff lady nearby that you could ask for help?"

"Yes, there was."

"Oh, Suzie, I think that's such a great idea that you had to ask her for help."

"Yes, I did. You simply are the very best to call me today. I love you to death."

"So, Suzie, when I first called you, I think you called me some silly name that hurt my feelings. Did you mean to do that?"

"Of course not. I just meant the walls are assholes, not you!"

"Oh. OK, Suzie. Thanks for explaining to me what you meant."

In her conversations, Suzie uses words and entire phrases that are familiar to her.

If the person speaking to her knows the almost scripted answers Suzie would like to hear, they are able to quickly get beyond the repetitive phrases and get into the real conversation with Suzie, about how her day has been or what she has planned for tomorrow.

It usually takes about ten minutes to go through the repetitive phrases of conversation and, in that time, Suzie will deal with her emotions surrounding her relationship with the person to whom she is talking and be ready to hear what they have to say.

If the person she is talking to would like to tell her about an upcoming event, they know to wait until Suzie's emotions have settled and then tell her the reason for the phone call.

Once Suzie has absorbed the information provided, she may ask the staff with her to write things down so that she will not forget.

Finishing up the conversation is also a process. This may be due to the fact that both beginning and closing the conversation is a transition and by using repetitive phrases, Suzie readies herself for it.

When a person Suzie talks to intends to end the conversation, they will say something like: "Well, Suzie thanks for talking to me today. I hope you will have a great night with your mom."

Suzie will then understand that you would like to finish talking to her. It may go like this:

"Thank you for calling me today."

"You are most welcome, Suzie. It was my pleasure to give you a hand when you were upset."

"Tomorrow is Wednesday, November 12. I don't want you to forget the date."

"Thank you, Suzie. I always appreciate it when you help me remember so I don't forget the date for my paperwork."

"Tomorrow is your garbage day. Don't forget to remind your kids to get your garbage can and recycling can out so that the garbage truck can empty them out."

"Oh, thank you, Suzie. I know we would have missed it without your help!"

"It's my pleasure to remind you! And please tell your husband that I will be helping him out tonight. I will get out of my chair by 9:00 p.m. and will go to bed early to get some rest for him."

"Oh, thank you, Suzie! We really appreciate it. You know my husband works long shifts and appreciates when you give him a hand."

"It's my pleasure! I can do anything for you and your family! I love your husband to death! Thank you again for calling me. Close your eyes now and I will sing you a goodnight song."

"Thank you, Suzie! I love when you sing to me!"

(Suzie sings the song.)

"Oh, Suzie, you simply have the loveliest singing voice. Thank you!"

"Close your eyes now!" (A quiet moment with sounds of kisses follows. Suzie kisses and rubs the receiver.)

"Thank you, Suzie. Hugs and kisses right back to you!"

"I love you to death!"

"I love you right back, Suzie! Bye now!"

"Bye! Thank you for calling me."

Conversations similar to the above examples usually happen with people who know Suzie well. However, when out in public, it is not known how people are going to respond to her. Suzie is a people person, and sometimes it has been impossible to stop her.

It is important to be diligent about helping Suzie choose something positive to share with the person she meets. The conversation needs to be facilitated by introducing Suzie to the person with whom she wants to talk, keeping in mind the one rule: Do not talk to strangers. Something may be said like: "Excuse me, sir, Suzie would like to talk to you for one moment," or "Suzie would like to tell you about her Christmas project."

Such introductions usually help the person feel more at ease with her. Other times, they may earn us a reprimand. Suzie might say, "Just stay out of it," showing that she wants to be independent in her interactions.

One goal is for her interactions to be natural. Brief chats with people in line at the cash register in the store or with dog walkers in the park while doing the "Adopt-a-Park" are culturally acceptable and are a great way to meet people in the community. The job of the staff is to help Suzie talk about positive things and help the person she meets to see Suzie for the great person that she is.

Getting around the negative initiation of the conversation is only one of the challenges while supporting Suzie in the community.

Every time she is out there in one of her jobs or volunteer opportunities, at social events, or shopping, it is important to try to troubleshoot before trouble happens.

One situation that comes to mind happened while delivering a Christmas hamper to the family Suzie sponsored through the Christmas Bureau. While Suzie was talking to the family, telling them excitedly about how she raised money by recycling cans and bottles all year long, Eliza, the person helping her, noticed that there was a bag of recyclables on the floor near the door. She signaled to me to be aware. At first I did not know what she meant, so she had to whisper to me that there was a bag at my feet. Once I realized what it was, I stood in such a way as to cover the bag from view. If Suzie had seen the bag, she would have immediately demanded a "donation" to her Christmas fund. If the family did not wish to donate their cans and bottles, this could easily have become a violent incident. Suzie may have become desperate to help them out by wanting to recycle their cans and bottles. In the midst of wanting so desperately to do something good and nice, Suzie would have lost her good judgement and turned this positive experience into a crisis.

If the situation had happened while visiting with someone who knew Suzie well, they could have either donated the bottles to her or turned the situation into an educational experience and explained to Suzie why she could not have the bottles, usually by deferring it to another time or offering something she could have done for them instead.

However, in public, when dealing with someone who does not realize the complexity of Suzie's understanding of the world, prevention has been the best strategy.

Another challenge is respecting socially acceptable boundaries. Suzie is a naturally affectionate person and likes to show her affection through hugs and kisses. This usually happens with people that she already knows, like her doctors, receptionists, or possibly someone she has just met who has done something to help her. Naturally, not everyone is comfortable with the hugs and kisses Suzie wishes to bestow upon them. As with anything else, just saying "no" will not do.

What is done instead is to help Suzie choose a more appropriate action. Offering Suzie an alternative, like demonstrating elbow kisses (rubbing one's elbow against the other person's elbow), can be enough to satisfy her need to show affection and alleviate any discomfort on the part of the other person. If the person with whom Suzie is interacting seems friendly enough, something could be said like: "Suzie would like to give you a hug. Is that OK?"

Another alternative might be to steer Suzie into deferring the hug or kiss by suggesting that she write a thank you card for the person to whom she would like to show her gratitude. Something could be said like: "I think this man has a sore throat. Can you think of something you could do for him?"

This has been done so often from an early age that Suzie will often come up with the idea herself. Since the number of people Suzie wants to write cards for is an ever-growing list, but time in her busy week limited, Suzie will usually forget about it.

One of the most critical times in Suzie's day is the 3:00 p.m. staff shift change. As mentioned before, transitions are difficult for Suzie. In the situation of daily shift change, transitioning from being supported by one person in the morning to a different person in the afternoon proves to be particularly challenging. Regardless of how comfortable Suzie is being supported by the staff on duty, and how stable the relationship, the excitement associated with the change usually takes over and results in colourful language being used by Suzie to greet the person coming in. All of us who work with Suzie know this well, and always make a conscious effort to get into the "Suzie mode" before we enter her home. We remain in that mode throughout the shift, not letting our guard down for a minute. All of our daily worries, family issues, and troubles have to be left outside Suzie's door so that we can support her successfully. It will likely end up being a lovely day for both Suzie and staff, if she receives our full, uninterrupted attention.

In preparation for shift changes, staff ladies often message each other so that the person coming on duty can be prepared for what they may encounter as they knock on Suzie's door.

If something negative happened during the previous shift, Suzie will want to tell the person coming in about it. It will frequently re-trigger negative emotions associated with the event, and may result in an escalation of Suzie's behaviour. In this case, staff coming on duty would do well to step back after knocking on the door and only come in once Suzie is ready to invite them. Standing too close to the door may result in being hit or spat on, if Suzie's emotions get the better of her. If the day went well, sometimes the excitement of telling about the good things may seem overwhelming to Suzie as well, and may present itself as escalation and seem negative. It takes experience to be able to tell the difference and assist Suzie in choosing appropriate words and actions.

Two Days in Seattle

BY SYLVIA DOANE

Walking from the hotel to a restaurant for dinner, Suzie accidentally bumped into a street person. "Watch where you are going, asshole," she said along with some additional colourful language. We moved her along quickly, trying to find and appeal to her soft and usually good-hearted nature. We tried to explain that the poor man had no coat to keep the rain off him. As well, he was probably hungry but had no money to buy food and likely had nowhere to sleep that night. Suzie listened quietly and attentively and when we thought she would be more understanding, she exclaimed, "So what? He's still an asshole!"

By trial and error, we learned how we could best deal with the ongoing behaviours. As Suzie had no impulse control, we could not "teach her how to behave." Again, trying to be proactive was always number one, but sometimes there was no alternative other than to just "take it," such as being repeatedly kicked by Suzie, who had thrown herself on the floor in a rage when being blocked from bolting behind a counter and frightening the person serving customers. It was necessary to protect others in the community, have them see that Suzie's staff would safely deal with situations, and subsequently not have Suzie live in isolation. We wanted Suzie to be accepted in her environment, wherever that was.

It was important to always be walking or standing in a balanced way when with Suzie, as she might suddenly give you a hard shove. She could knock you down or into a wall if you were not ready at all times. Spitting in a staff member's face was degrading and horrific to some, ignored and with no reaction by others. Learning to "not take it personally" was often a challenge and a learning experience for the staff team. On the inside, you were often taking it personally, although not showing it.

It was discouraging to have had a fine day at school, an uneventful drive home, and then suddenly be a target for thrown silverware as soon as you both went into the house. We knew it was just shift change, but somehow it was always such a disappointment. The saving grace was the closeness and kindness of the on-shift staff and the never-ending search for a solution. When there was success for the after-school shift change, it was a great step forward as well as a cause for celebration!

When going into the changing room before and after swimming, we had to be extra fast getting our clothes or swimsuit on as Suzie would try to bolt out while stark naked. It was best if we were not blocking her exit also unclothed.

In the early years, the incidents of aggression and violence occurred many times each day, so writing incident reports for each of them was just not possible or expected. One of the situations staff found themselves in included the following report:

"I had stopped in a parking lot in Ladner when Suzie began kicking the dashboard of my car. I could not continue our journey home until she had calmed down. I was restraining her right leg with my hand (gently but firmly) and talking to her. Suzie was screaming, 'Get your hands off me, and don't touch me!' I decided I would turn back to Gateway where I would have some help and would get Suzie back in a good place before we began the drive home yet again. In the meantime, some well-meaning bystanders called the police, thinking Suzie was being hurt by me. The police arrived at Gateway about the same time we did, said they were pretty sure it was a behavioural episode with a Gateway student, asked me if I was OK, saw that we were back in the school, and left. In a short while, Suzie was ready to go home and we had a fine and uneventful drive. We had over thirty kilometres drive to get home, and I could not safely travel with her being so aggressive."

In the later years, as strategies were developed, the incidents became less frequent, so they would usually be recorded in the form of incident reports.

Here are some examples of behavioural incidents in recent years:

APRIL 23, 2013

Suzie woke up at 7:10 a.m. with a lovely whistle for me. I went into Suzie's room and socialized while she had a chance to wake up enough to get off her bed.

Suzie was in a good mood and was even laughing with me. Once fully awake, Suzie went into the kitchen to feed her cat, Lucy, as she does every morning.

She also unplugged her car cell phone, which was plugged in after I did the car check, and it was still dead. Suzie whistled for me to hold the phone for her while she got it unplugged.

When it didn't light up the way it usually does, Suzie said: "I don't know what's going on with my stupid car phone."

So I told Suzie that maybe it wasn't finished being charged, and that maybe we should leave it plugged in until after breakfast.

Suzie then shook her head and said, "No, you liar. It's not allowed to be plugged in no more. I'm going to have to give you a dig in the eye."

Suzie, while saying that, picked up the scissors that were on the counter and threw them towards me. I saw her reach for them and moved out of the way. They hit the wall by the front door. I then picked them up so as not to let Suzie get hold of them again and moved myself down the hall into the living room.

As Suzie kept saying, "Give me the scissors back," I paced around the furniture while trying to de-escalate the situation. Suzie then said, "I'm going to have to punch you."

She picked up a picture that was on the ledge by the stairs and threw it towards me. I picked it up and said to Suzie: "Please let me help you, Suzie. How about we call the supervisor to give us a hand?"

Suzie then walked away and sat in her chair. I used my cell phone to call the supervisor.

I passed the phone to Suzie. The supervisor continued to de-escalate Suzie's behaviour while I went and got the PRN (medication) for her.

FEBRUARY 9, 2014

I took Suzie to a movie. We were a bit late for the movie. As we were walking to the theatre, a staff member was taking tickets. I handed him the tickets as I was holding them. When I did this, Suzie got upset. She told me she wanted to hand them to him. I asked him to hand them back to her. She then spit on my face, pulled my hair, and kicked me. I re-directed Suzie and got her down to a quieter state.

We went in and watched the movie. While watching the movie, Suzie and I conversed about the movie; she seemed happy and enjoyed the movie. After the movie, while we were walking out, Suzie informed me she was still mad at me and that she didn't want me to work with her anymore.

She then told me she was going to punch me in the face and that I deserved that. She then grabbed my hair very hard, almost pulling me right over; she spat on me, kicked me, hit me, and swore at me. I moved away from her, but she chased me, swearing at me. People attending the movie asked me if I was OK and if I wanted them to call the police.

I asked them not to call the police and informed them I was calling my backup manager.

JUNE 3, 2015

Suzie and staff member were out cleaning the complex. Suzie was almost finished when she saw two young boys playing with Nerf guns along the route that we were headed down to finish our circuit. There were a few Nerf darts lying on the street. Suzie saw one in front of her; she picked it up with her garbage picker and put it in the garbage cart.

One of the boys saw this. He ran over to the cart and stuck his hand in to fish the dart out. Suzie swung her garbage picker at him and hit him on the side of the head. The boy crouched on the ground, holding his head and started crying. Staff rushed over to see if he was OK. The boy didn't reply then ran to his friend's house just a few houses down from where the incident occurred.

When the friend's mom came outside, staff member explained what had happened. The mom offered to walk the boy to his own house and tell his parents what had happened.

Staff member stayed behind with Suzie and called the supervisor to inform her of the situation and to talk to Suzie, who was still agitated, saying that the boys were not allowed to play outside.

Suzie spat at staff and threatened to hit them with the picker as well. After speaking with the supervisor for a few minutes, Suzie calmed down and agreed to go home.

At this point, the friend's mom came back and let staff know that the boy was OK and that his mom had been very understanding about the incident.

OCTOBER 14, 2015

Suzie, Eliza, and I were in the front hallway chatting about the day at shift change. Suzie was hanging up her coat in the hall closet.

I asked her if she still had enough energy to do her journal after her busy day. Her mood instantly turned and she told me to shut up and lunged at me to hit me with the hanger she was holding.

Eliza caught the hanger and asked Suzie to stop. Suzie turned her anger on Eliza and started hitting her on the head and body with the hanger and spitting on her. Eliza shielded me from Suzie and tried to block the blows, telling Suzie it was OK and suggesting that she go and sit down in her chair. Suzie and Eliza moved down the hallway while Suzie continued to hit Eliza, spit at her, threaten her, and swear.

Suzie threw the hanger towards us then sat down in her chair. Eliza started talking to her and calming her down. I prepared 1.5 mg Ativan as per the agitation protocol and messaged the supervisor to let her know what had happened. The supervisor left immediately to come over to Suzi's. Meanwhile, we continued to talk with Suzie. She eventually took her medication (Ativan) at 4:30 p.m. and the supervisor arrived shortly afterwards.

Suzie displayed anger at Eliza again with name-calling and spitting when Aga (the supervisor) arrived. Suzie struggled with making a good choice, continuing to use rude language towards everybody. The three of us debriefed with Suzie until 5:00 p.m. when she was feeling calm again.

FEBRUARY 7, 2016

Today Suzie had a late start in the day, waking up around 3:00 p.m. After getting up and changing and getting weighed, she sat down in her chair and worked on her journal and prayers while having her breakfast.

I waited until after 7:00 p.m. to have her dinner prepared and put on the table. She was upset and told me she wanted to listen to music.

I told her that that was fine if she wanted to listen to some music first. She became agitated throughout the day and asked to speak to staff member Ami on the phone twice throughout the shift.

While I was helping her to get her music items ready, she told me that I was being stubborn. After a little while the situation escalated and she became very upset, telling me I needed to get Ami on the phone.

I asked her if she wanted to talk about what was upsetting her. Suzie told me that she was going to have to "punch me in the stomach really hard" and that I deserved to be kicked and hit.

I told her I didn't want to keep her waiting and that I had just prepared dinner, but we could wait until she was ready. After speaking to Ami, she felt better, but around 7:30 p.m. while helping to put her music away she became agitated again.

She tried to hit me, and I moved out of the way. She attempted to hit me a few more times while telling me to pick up a CD case. I dodged her each time.

When I moved out of the way, she became even more agitated that I wasn't letting her hit me. I asked her why she felt like she needed to hit me and that we could talk about what was bugging her.

She got up and began to hit me and pull my arm. For about a minute or two she continued to hit, shove, punch, kick, and pull at me and my arm.

I kept my arm up defensively to try to lessen the shock of her hitting anywhere else on my body. I continued to back away until I was at the front door.

She grabbed my left hand, which I was using to defend myself, and bent four of my fingers back.

I kept speaking to her and explained that she was hurting me and that we could call Ami again if she needed to speak to someone and do the relaxing breaths.

She continued to tell me to shut up and that I deserved this. While leaning back a little to punch me again, she lost her balance on her foot and she backed up into the wall and fell down in a sitting position.

I could not fully check her to see if she was OK, but the fall didn't seem as though it had hurt her badly.

I went into the kitchen, grabbed my cell phone and keys, and called another staff member from my cell phone at 7:37 p.m. to tell them what was happening. Ami then called two other staff members.

Staff member Eliza was able to come and help me in a matter of a few minutes. While Eliza was on her way, Suzie was still sitting on the floor, telling me I needed to help her up while still using rude language.

I stood outside the door but kept it slightly open while staying on the phone. She helped herself up onto her knees and pushed the door closed.

I continued to speak to her and told her that Eliza is on her way to help. She continued to push the door closed before getting on her knees and shutting the door before locking it.

Eliza came, comforted me, and used her key to get inside and help Suzie.

SUZIE'S USE OF SPOKEN LANGUAGE

One of the most recognizable characteristics of Suzie is her speech. Everything she says is highly emphasized, either positive or negative. Suzie developed her own language that is quickly learned and adopted by those close to her. When she wants to tell someone something that in her opinion is most exciting, she will often say: "You will be so shocked!" or "You just won't believe it!" with utmost expression.

When Suzie has a seizure and later wants to tell someone about it, she will usually say "Those silly seizures drive me banana nuts crazy."

Should a staff member supporting Suzie be in the kitchen preparing a meal, Suzie might call them from the living room by shouting: "Get your big fat ass over here right now!" and quickly add: "if you have a moment, Your Majesty!"

When we share something good that happened to us, Suzie will often say: "Oh, you lucky you, spoiled rotten you are!"

When Suzie wants to express the "not so keen" opinion of someone, she may invent all sorts of names for them, preferably ending with a head of some sort: fart head, skunk-head, giraffe-head, monster-mash-head—you name it!

If you want to interject while Suzie is having a conversation with someone, you may be told to, "Shut up your ass and stay out of it, you pervert!" as if you were going to break-up an argument or something.

Suzie's terms of endearment may vary greatly, depending on her mood and level of excitement. She may be calling her "blasted Mom" or "dear God-blessed Mom," or her staff a "stupid jerk" or "my sweet blueberry buttercup." She may be kindly asking, "May I have your attention for a moment, please," or "You better listen to me right now, you fricken idiot!" You just never know which one it might be! Gotta love her!

CHAPTER 6

History of Suzie's Volunteering and Employment

BY AGA KARST

From the time Suzie was really young, people close to her found out very quickly that she loves to help. She will do anything you ask of her if she knows that it will help somebody out. So it only made sense to channel all the wonderful energy she had into something positive and constructive. It was always very important to Suzie's family that she would give back to the community that gives so much to her. We all want Suzie to be known for the wonderful, giving person she is rather than for her "behaviour."

At this point, it is probably difficult to remember what Suzie's very first volunteer involvement was, but we know for sure that cleaning toys at a local Health Unit was one of the first.

Sylvia fondly recalls:

"It all began very innocently when Suzie was at the Health Unit to have a vaccination shot. While there, Suzie talked about looking for a volunteer job. They asked her if she would like to come in once a week and wash the toys that were there for children to play with. Suzie did such a good job, she kept the job for three years. For nearly two hours every Wednesday, Suzie would get all the cleaning supplies ready and we would wash the books and toys together. Suzie especially enjoyed the times when little ones arrived for their appointments.

Suzie would sing the children songs and let the children and their parents know which of the toys were clean and could be played with. Suzie also washed down all the large toy slides, climbing equipment, and so on.

One of the occupational therapists was so impressed with Suzie's work and attitude that she asked Suzie to come to her work area and clean the toys she had as well. Suzie so likes to help people, and she would respond, "Of course I will." Suzie was known and loved by all she worked with as well as those who came to the H.U. for services. Each week when Suzie finished her work, she would walk down to the Office Manager's office and tell her what she had cleaned that day. Suzie had a wonderful relationship with the manager. Each week she would give Suzie pencils, stickers, and so on as a way of thanking her, and Suzie always looked forward to this time.

Suzie "retired" after nearly three years. To this day people still stop her and say "I remember you from the health unit!"

This work experience required constant attention, support, and role modelling from the staff that was there with Suzie. It was important to always be on the alert for any "triggers" that might have upset Suzie. Always being vigilant to any possible distraction and being able to quickly re-direct Suzie on the right path helped ensure a positive experience for everyone.

This was most likely the first experience, which started a long and intense volunteering path for Suzie. Suzie volunteered for about eight years at St. Catherine's School Library. This volunteer opportunity had to be very carefully planned and organized, as lots of young children don't exactly create a "controlled environment" where everything would be predictable for Suzie. Before Suzie was able to go in, a meeting was called for all teachers and teacher's assistants who worked in the school. They were briefed on who Suzie was and what she would be doing (shelving books in the library). Teachers were given strategies on how to respond to Suzie if she approached them in the hallway, and what to say to her to assure positive interaction. An added bonus was that one of Suzie's residential support staff also worked at the school as a teacher's assistant. This became a valuable safeguard. If Suzie got upset with the support staff that was with her, this other staff could be called in for backup. Teachers in the school understood this and didn't mind letting her attend to Suzie when needed. Routine and timing of these visits was carefully developed to assure success.

Mrs. Piper, the head librarian, recalls:

"Suzie always called me Mrs. Piper. If I told her that she could call me Maria, Suzie would say "I know that, but I like to call you Mrs. Piper." She would always notice when my ring was crooked and insisted on fixing it for me. When she worked, she always knew where the books went, especially the ones from the younger children's section. Suzie didn't like it if somebody tried to touch or take books off the shelves while she was working. She would say, "You are not allowed to touch them." After completing her work, Suzie would always write me a note. She called the kids "scally waggles" or "monkey's uncles" in her note. I feel so sad that I didn't save the notes Suzie wrote me."

Suzie was very proud of her job at the library.

One of the very important projects for Suzie is Adopt-a-Park. Suzie goes to a local park in her community and picks up litter. She has done this for many years and is still committed to it. She has later obtained paid employment in her townhouse strata (picking up litter in the complex) based on this volunteer experience. Suzie is very proud, and rightly so, of keeping the community clean and safe "for children and pets." If she meets neighbours or friends as she cleans, she tells them what she does and loves receiving acclamation for her work. She also proudly tells those she meets that there is a board up in the park that says "Park adopted by Suzie Bailey—and that's me," she adds with a huge smile.

Suzie delivered internal mail for LACL for many years. She loved to help the administrative staff out at head office by taking mail to all the different programs out there.

Based on this experience, Suzie later volunteered at Meals on Wheels, another volunteer-operated service which benefitted from years of Suzie's commitment. Every Friday, Suzie would deliver hot meals to seniors in her community. She developed friendships with many of them, and would often stay for a chat or short visit when delivering their meal.

St. Nicolas Church is a local Catholic parish where Suzie went every week for over eight years to fold bulletins. "You will not believe it! I folded up over three hundred bulletins!" she proudly told everyone she saw after completing her weekly assignment. The social and spiritual aspects of this particular project were just as important to Suzie as the work itself. She always loved a visit with the church secretary or one of the priests. After the work was done, Suzie would go into the sanctuary and talk to God. She prayed out loud, and her prayers would without fail bring a tear to the eye of her support staff as well as to anyone who may have been having their own quiet time in the sanctuary. You couldn't help but cry when Suzie faithfully brought all her requests to the Lord, committing her loved ones in prayer to God, as well as strangers she had never met. If Suzie saw someone getting into a car accident on the road, she would always remember to pray "for those poor people who got hurt." You also couldn't help but laugh when Suzie finished her prayers with a loud "OK God, Jesus, and Lord, you have a fun weekend up there in heaven and enjoy your days off! I will see you again next Friday!"

Seasonally, Suzie would also get involved in fundraising for various charitable organizations. She participated in door-to-door fundraising for the Heart and Stroke Foundation, helped with bottle drives for the Special Olympics, rang the bell for the Salvation Army Kettle campaign, and completed a five kilometre "Poker Walk" for the BC Epilepsy Society.

Perhaps the project that brings the most joy to Suzie, her family, and staff is her bottle recycling project. It started small and grew to enormous proportions. During one of Suzie's annual planning meetings, her support staff Vonna came up with the idea that Suzie could recycle all her cans and bottles, put money into a special account, and then use it for a charitable purpose. Suzie instantly loved this idea! She would go to the bottle depot every week and tirelessly sort her cans and bottles into trays. It was then decided that Suzie would adopt a local family through the Christmas Bureau. By the time Christmas came about, Suzie had almost three hundred dollars saved and she sponsored a single mother in her community.

From that point on, the project snowballed. As people heard about what Suzie does, they would donate their cans and bottles to her. For a couple of years, she even received collected cans and bottles from an entire apartment building where one of her staff ladies lives. Each year Suzie sponsored a bigger family, some years spending as much as eight hundred dollars. The recipients of Suzie's hampers would have tears in their eyes when she told them how she saved money to help them out. We on Suzie's support team believe that this project may have had the biggest impact on the community of all her volunteer involvement. People she helped out would often say, "Wow, if she can do it, I can too!" Some of the years, Suzie's donations of bottles and cans were so large that we could barely keep up recycling them. Some years she collected so much money, she was able to make hampers for two or three families if it were physically possible to do that much shopping! Consulting with her family and staff team, Suzie would decide who else she wanted to help in the years that she had extra money. For two years in a row she made donations to the "Smile Train" foundation and sponsored children in third world countries to receive funds for cleft palate surgeries. One year, right around Christmas time, LACL was having a fundraiser for a local family, and Suzie donated $500 to them.

Suzie was featured in the local and provincial papers regarding her Christmas Bureau project.

Suzie Bailey Filling the Candy Machine

Red Apple Grand Opening in Florenceville

Gifted with compassion

Suzie Bailey has worked all year so that two Langley families will have a merry Christmas

MONIQUE TAMMINGA

Walnut Grove's Suzie Bailey has an acquired brain injury but that doesn't stop her from volunteering and supporting worthy local charities. This year, Suzie adopted two families through the Langley Christmas Bureau and has been collecting bottles and cans to help ensure they have good food to eat and presents under the tree.

"People shouldn't feel sorry for Suzie. She is very capable and has a fulfilling life."
Vonna Nugter

Credits to Monique Tamminga, Langley Times Reporter and to Black Press

Yuletide spirit

Suzie's giving grows each yea

Suzie Bailey loves Christmas and devotes a great deal of time to helping others have special holiday memories.

A Walnut Grove woman loves being one of Santa's elves and works to help more people each Christmas.

$620.23 – To reach that total, Suzie Bailey had to cash in a whole lot of cans and bottles, five, 10, and 20 cents at a time.

The 28-year-old Walnut Grove woman, who despite an acquired brain injury, strives to maintain an active life, devotes a great deal of energy to collecting, sorting, and cashing-in containers for deposit for one reason.

She wants to provide people with special Christmas memories.

"I love Christmas," Bailey said.

Her holidays are spent surrounded by loved ones.

"I'm looking forward at Christmas to spending time with my dear mom and dad and (extended family)," she said.

In the days leading up to the holiday, she's also busy cooking treats and making crafts as gifts.

But she's also active around the community. Her housing complex hires her to do cleaning and allows her to have the recyclable containers.

The Otter Coop, where she also works, and the club where she paints also contributed cans and bottles.

She did this fundraising last year and sponsored a grandmother and grandson through the Langley Christmas Bureau. This year she upped her game and is sponsoring two families.

Bailey and her care workers went shopping to pick out items specifically for the two families. The bundles were wrapped and earlier this month, taken to the families, which she looks forward to meeting each year.

One family is a mom with a young son, and the other is a mom with a one-year-old girl and a three-year-old girl with cerebral palsy.

"We've contacted the moms and they're super excited to meet Suzie," said Vonna Nugget, pa... Bailey's care team.

One of the moms is saving up her cans and bottles for Bailey.

"For next year we h... bigger plan," Nugget...

Bailey is looking to... a large family and als... contribute to Variety...

Credits to Heather Colpitts, Langley Advance Reporter

After she appeared in *The Province* newspaper, a television station called. They wanted to make a TV report on her and name her "the volunteer queen." Too bad their timeline was so short that it didn't give Suzie enough time to plan on meeting with the reporters in the midst of her super busy life! She was busy helping others and didn't have time to become famous!

Naturally, so many volunteer commitments helped Suzie gain valuable experience, which helped her secure some paid employment to supplement her pension and help raise her standard of living. It was also a great boost in self-esteem to receive a cheque written in her name for work well done!

Initially, Suzie received a placement for "work experience" at KFC. She would go there before the store opened for lunch. Her job was to prepare and date salads in individual containers. Macaroni, potato, coleslaw—Suzie could make them all! Even though this particular work experience didn't lead to employment, it was a great way to learn what having a job involves, develop new skills, and boost her resume.

From this placement, Suzie moved on to a real paid job. Her first one was at Otter Co-op. Suzie would go to their office to help shred paper once per week. On good days she would spend most of the day there, taking a break for lunch. She would shred two to three boxes in a big shredder machine placed in the walk-in storage closet. Suzie loved getting to know her co-workers by name, and had a routine of talking to several people before starting work and upon completing it. She couldn't help but tell them how she was helping them out. Of course, receiving a cheque in her name every two weeks was also something to brag about! Suzie maintained that employment for eight years. She only "retired" when the amount of shredding needed by the Co-op was more than anyone could handle by hand, and a large shredding company was hired. However, all Suzie's co-workers there were really sad to see her go, and they held a big party for her to say good-bye.

Suzie's second job was litter pick-up in her complex. She was hired by Woodbridge Strata after she applied there, listing her volunteer experience with "Adopt-a-Park" on her resume. Suzie has been cleaning her complex for about five years now and is still going strong. Suzie had the opportunity to meet a lot of her neighbours while on the job. She loves chatting with everyone she meets while she cleans, whether they are out walking their dogs or working in their gardens. Sometimes she "reminds them" to turn off their outside lights during the day, sometimes she tells them about her work, and sometimes she simply wishes them a great day. Without fail, Suzie brings a smile to everyone's face, and they really value her work.

Another job Suzie held for a number of years was working for Gibbs Nursery. This is how Sylvia went about getting Suzie employed there:

Once when buying flowers at Gibb's Nurseries, I spoke to one of the owners about the possibility of employing Suzie, the young woman with whom I worked. I talked about Suzie wanting to work and earn money and how important this would make her feel. She said she would have to talk to her husband, since he was the one who managed the outdoor area of the nursery. We agreed that I would check back with her in a week.

When I returned the following week, she admitted to forgetting about our conversation. I asked if she would mind if I spoke to her husband directly, and she said that that would be fine. I found him and explained to him that I had already spoken to his wife, and I asked him how he would feel about employing Suzie to work one hour a day for one day a week for minimum wage. I also confirmed that I would always be with Suzie to support her in her efforts. He agreed that it was fine with him if his wife was agreeable to the idea. I went back inside and confirmed the agreement.

A few days later, I took Suzie in and introduced her to the owners. They bonded immediately, and Suzie started work the following Friday.

During the nearly three years that Suzie was employed at the nursery, she hardly missed a day of work. She was always happy, going about her job of dead-heading flowers and keeping the greenhouse tidy. She built up a strong and caring rapport with all of the staff there and enjoyed a great experience week in and week out.

Sadly, one day, the owners announced that they had sold the property for development and would be retiring. Suzie was one of the last employees to be given her notice, and it was not without sad goodbyes and lots of tears all around. Upon her leaving, Suzie received an excellent letter of reference from Sharon. All in all, this was a wonderful learning experience for everyone involved, including Suzie, myself, the nursery staff, and many of the customers at Gibb's Nurseries.

Gibbs Nurseryland & Florist

7950 200TH STREET, LANGLEY, BC

SEPTEMBER 30, 2012

To Whom It May Concern,

Suzie Bailey has worked for us for the last three years throughout all seasons. She is bright and energetic, endearing herself to us all. She has completed tasks such as packet stapling, box making, plant dead-heading, watering, and water-tube filling. She also really enjoyed helping with pricing new stock. She completed these tasks with great attention to detail and a strong work ethic and often interacted with our customers, never failing to bring a smile to their faces. With the closure of our business, we will miss Suzie's Friday visits and hope that she is successful in finding another place to work. She certainly has our support and recommendation. Suzie would be an asset to anybody's business.

Sincerely,

Work experience at KFC.

"The kids at school called me Ms. Bailey, the librarian. I put the books where they belong."

"I love helping out at St. Nicolas church. I folded up 300 bulletins!"

"I recycle cans and bottles at save money to help families in need at Christmas time."

Based on this letter and the experience she has gained, Suzie was able to secure a similar job at Art's Nursery. It is seasonal work, but Suzie has maintained it for several seasons now. Again, all those who come into contact with Suzie during her work quickly come to love her and will often stop by Suzie's work station to say hello.

One would think that all the volunteer involvement and paid jobs would tire her out, but no, not Suzie! She actually has enough energy to pursue self-employment as well. She must be taking after her dad with her entrepreneurial spirit! With some "seed money" entrusted to her by her parents, Suzie started her own small business of candy vending machines. Her first machine was placed at Day and Ross trucking company (they featured an article about Suzie in their newsletter!), her second one is at Valley Driving School in Langley, and her newest machine was recently installed at a newly-opened location of Valley Driving School in Aldergrove. Suzie loves to go to service her machines every week. Collecting the coins, refilling and cleaning the machines is just as valuable to her as talking to people she sees there. In her mind, she is just helping all those awesome workers and students by providing candy for them.

With money saved in the bank, Suzie is able to help pay for her annual vacations.

This business is ideal for Suzie, as it provides her with flexibility to work around her busy schedule rather than being tied down to specific work hours.

Suzie's entrepreneurial spirit doesn't end there. She also takes on odd jobs—particularly, she is known as "The Cat Lady." For a small fee and some chocolate, she will look after her friends' or relative's cats if they are away on vacation, or busy with work or school. She will feed them, make sure they have water, and even deal with a stinky cat litter box! Leave it to Suzie to take the garbage out with her as well, so that you don't have to return to a stinky garbage can!

CHAPTER 7

Vacations

BY SYLVIA DOANE

In her young adult life, Suzie has travelled to Osoyoos, B.C., with her friend for a few days of sun and R & R. She travelled to Victoria to attend a Rod Stewart concert and spent some time by the ocean at the Tigh-Na-Mara Resort on Vancouver Island. Suzie really enjoyed a cruise to Mexico and passed the time before and after the cruise relaxing in San Diego, where she enjoyed a few special meals, including treating her mom to a wonderful brunch on Mother's Day. Suzie has also enjoyed a shopping and sight-seeing weekend in Seattle.

Suzie's most enjoyable getaways are still her regular trips to Las Vegas. This is her favourite place to hang out. Suzie has travelled to Las Vegas several times and experienced a number of events, including attending live concerts to see Céline Dion, Elton John, Shania Twain, and Cirque du Soleil shows that showcased The Beatles and Elvis. She might spend a few dollars on the slot machines, but her real enjoyment comes from seeing the shows and, of course, the "shop-till-you-drop" experiences she always looks forward to with great excitement. It is worth mentioning that each time Suzie earned and saved at least half of the expenses of these vacations, with her parents contributing the balance of the costs.

One travel experience Suzie truly enjoyed was travelling to Halifax, Nova Scotia, to be a bridesmaid at her sister Sam's wedding. Suzie made the long flights there and back like a real "trooper" and thoroughly enjoyed the whole wedding celebration.

THE WEDDING

By Sylvia Doane

Wednesday, February 13, 2013

Day 1: 6:15 a.m.

Suzie was ready to go when Wayne (my husband) and I arrived to pick her up. She was in a super good mood on the drive to the airport. She did all the talking.

Suzie did a great job of checking in and walking through the security check-in like an old pro. We had time for breakfast before boarding.

The flight was just lovely. Suzie kept herself busy with her iPod and colouring for the nearly five hour trip to Montreal, where we changed planes to continue on to Halifax. I was so pleased with how efficient the airline was with connecting flights and wheelchair assistance. The pleasant staff also made it enjoyable. There was one instance while changing planes in Montreal that I was the one left pushing the wheelchair. It was a steep ramp down to the plane. I was having some trouble holding the wheelchair back from taking off when one of the handles came off. Then we really took off and picked up speed. Left with only one handle, I was able to steer it into the wall. After some crunching, I was able to bring the chair to a stop. My heart was pounding like hell, and Suzie was laughing her head off, which set me off as well. The noise brought some airline fellow to our aid. "Are you hurt? Are you OK?" he asked. We were laughing so much it was hard to answer back but yes, we were OK! Just another little moment to remember.

Suzie's dad, Bob, met us at the airport, and Suzie was very excited to see him, greeting him with some of her colourful phrases. She quickly settled down to her best self on the drive to the hotel. Awaiting Suzie in the hotel room was a Valentine card and gifts from her mom and dad, adding great excitement. We went down to the restaurant for dinner with her parents. Suzie could not wait to give her mom, Cathy, her birthday gift and sing her "Happy Birthday." Cathy was so touched.

Back in the room, Suzie asked if she could listen to music and colour for a while. I asked for her to set a time to turn the lights out (it was now 11:00 p.m.), and we agreed on midnight. This allowed me some time to unpack a few things. After an eighteen-hour day, lights were out at the stroke of midnight. Sweet dreams and no seizures.

Day 2:
Suzie was awake by 8:30 a.m. I made coffee and hot chocolate. A little later, Suzie's mom came up for a visit while I went downstairs and brought breakfast up to the room. Suzie enjoys having breakfast in bed. I showered while Suzie listened to music. I don't know how, but I seemed to have a second shower while assisting Suzie with her shower. We had lunch at the hotel while everyone there got to know Suzie.

Suzie's mom and dad picked us up at 5:00 p.m. for the rehearsal dinner. Suzie walked right into the kitchen ahead of us and was busy socializing before we could catch up to her. The lobster dinner was delicious, and Suzie was wonderful at keeping it together with eight children at the table. Their average age was four years, and they were charging around and hollering their cute little heads off, just being kids.

We met most of the family at the dinner and then it was a wild rush to get to the church. The rehearsal was hilarious with the padre trying to organize everyone, eight kids "flying from the rafters," and Suzie wanting to tell the padre to "shut-up his face and stop talking." Suzie also argued that she should be walking her sister down the aisle and not Sam's "blasted father." Cathy and I had a few giggles over the whole zoo-like atmosphere.

Then it was hot chocolate and lights out at 11:30 p.m. Suzie had a few focal seizures in the early morning and went right back to sleep.

Day 3: The Wedding Day

The morning routine was the same as the day before, with a visit from Mom chatting about the day before. Cathy made hair appointments for Suzie and me. We travelled back and forth by cab. I believe Suzie was the highlight of everyone's day at the beauty parlour. Everyone knew why she was getting her hair done. I have no doubt Suzie left a few pleasant memories there. After a quick lunch, it was time to get beautified. Suzie looked stunning. When we arrived downstairs, most of the hotel staff seemed to be there telling her how gorgeous she looked. Suzie loved every minute of it. Then we took the taxi to the chapel.

Suzie was still determined that she was going to tell the padre to "shut-up your face" and that Dad was not going to walk Sam down the aisle! We tried this: "Suzie, I promise the padre will not talk too much today. He will just quickly marry Martel and Sam. Can you handle that?"

"Yes, I can."

"Suzie, Dad walked Sarah down the aisle, and now it's Sam's turn, and if you get married, he will walk you down the aisle too, because that is what dads do when their daughters get married. What do you think?"

"OK. He can do it this time."

What a wonderfully proud moment it was watching Suzie walk down the aisle (holding her flowers) by herself and then stand beside the rest of the bridesmaids. It was a lovely, lovely wedding.

After the reception and speeches, the emcee asked if anyone would like to come to the microphone and say a few words. I asked Suzie if she would like to and she said, "I sure would." I walked her to the podium, and Martel's brother handed her the microphone while I went back to sit down. You could hear a pin drop!

Suzie said, "Congratulations, Martel and Sam. You spoiled rotten wife, you are." She then sang "Twinkle, Twinkle Little Star" and said congratulations again. She handed the microphone back and walked to our table and sat down. Sam had tears in her eyes and was so moved, as was everyone else. It was a very special moment for Suzie to shine, and shine she did!

She happily danced the rest of the night away. When we finally returned to the hotel, Suzie sat on the bed and said that her feet hurt. Removing her shoes, I saw a painful looking red ridge across each foot. It just goes to show what we women will endure for a little glamour. You know the saying, "Beauty knows no pain."

Day 4:
It was a lazy morning. We had breakfast in the room, showered, and packed. Lunch was with Mom and Dad, and after that we went off to the airport with a quick stop at Tim Horton's. Then off on a long but lovely flight home. Wayne met us at the airport, and we headed home. Suzie talked to Wayne non-stop about the wedding and what a good time she had had in Halifax at the wedding and with her family. We were at Suzie's home at 10:00 p.m. Just before I was leaving, Suzie put her arms around me and said, "Sylvia, thank you for taking me to Halifax. I love you." How sweet it is to be loved by you, Suzie!

All in all, it was a great, memorable trip.

Going to the movies

BY SYLVIA DOANE

Going to the movies with Suzie provides a whole secondary kind of movie entertainment. In addition to the actors and action happening on the big screen, there is another monologue going on, thanks to Suzie and her views and interpretations of what is happening. If she disagrees with what is going on or what one actor is saying to another actor, she will suggest that that actor, "Shut your mouth, it's none of your business," in a loud and no uncertain manner. As well, if someone on screen is not being nice to a child or a woman, or being mean to an animal, out comes, "You're being an ass-xxxx and deserve to be kicked and put in jail," much to the amusement of most everyone within earshot.

Oh yes, she gets into the movie and participates in an enthusiastic and boisterous manner. Yes, she does! Maybe being a director or producer could be in her future?

"Having a laugh with my friend, Wade."

At River Rock Casino to see the Jacksons.

"Dad brought along champagne for my 30th Birthday"

"I kicked everybody's butt at bowling, haha!"

Our last visit with Dr. Farrell.

"My Mom and I always enjoy our visits with Dr. Hurwitz!"

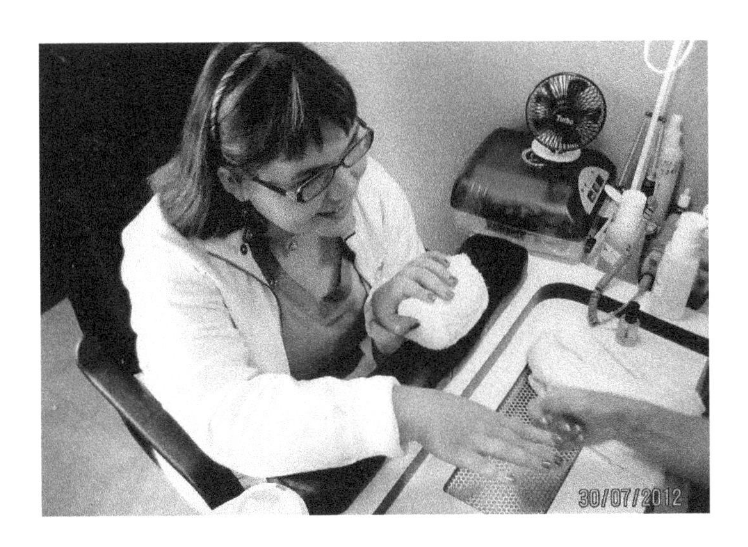

"I so enjoy my monthly manicure!"

Relaxing before the concert in Vegas.

"So excited to see Céline Dion!"

"I want to go see Rod Stewart next!"

"I'm off to the ball!" LACL's 50th Anniversary Gala

"Working on my monthly calendar."

CHAPTER 8

Suzie's Support Staff

BY AGA KARST

Those of us who are close to Suzie and her family often reflect on what it is that helps Suzie be successful. What is it that makes her beat the odds? What caused the tremendous impact she is having on her community, despite the challenges she lives with every day? The sheer number of seizures she faces every day would incapacitate almost anyone else. Brain injury and subsequent behaviour would likely lead a different person to being institutionalized, heavily medicated, or living in seclusion. The number of medications she takes (mostly to treat medical issues, not behaviour), producing side effects of drowsiness, fatigue, lack of balance, and other symptoms, would likely cause a less determined person to become homebound. However, Suzie continuously beats the odds and is able to live a successful life.

Most of all, it is Suzie herself. She has a heart of gold and is motivated solely by her need to help others. She has a zest for life and a drive for independence that has amazed many along the way. Those of us who are close to her realize it is due to the fact that she wants to help. Suzie does not do things because she feels like it. She does not do them to satisfy her needs or wants. She does not do them because that is what society expects (although it may seem that way when she shares what she did with every person she meets). She simply does things because she wants to help.

Secondly, Suzie is nurtured by her wholesome, loving family who cherish and love her. They are always involved and present. Living very full lives themselves, they don't expect anything less of Suzie. Setting the bar high, they create an environment for success. They are most supportive whenever Suzie herself or her team comes up with new ideas on how she can contribute to her community or bless those close to her. We have never heard any of Suzie's family express worry or concern that Suzie might somehow hurt herself or put herself at risk as a result of her generosity. What may raise an eyebrow or even cause a panic attack in some more protective parents or siblings puts a smile on the faces of Suzie's parents and sisters and excitedly they say, "Yes, she can!"

Suzie's medical team has always been nothing short of excellent. Bob and Cathy Bailey often reflect on how blessed they have been by the excellent medical care she received. Children's Hospital in Vancouver tops the list, with Dr. Farrell always being involved. Later, Dr. Hurwitz of UBC joined the team. Perhaps the caring and personal approach they all offered was equally important to the medical expertise they had.

From the moment BettyAnne Batt and Dan Collins of the Langley Association of Community Living came on the scene, they saw the desperate need this child and her family had. They advocated tirelessly for the right supports for them. What started off as a respite agreement evolved into a life-long support system. Securing funding has always been a challenge with the limited resources the health ministry could provide. Suzie just did not fit into any of the categories the ministry had, with her needs highly exceeding those of others. At one point, Suzie was named "the most expensive child in the province." However, this did not stop Dan from negotiating contracts that Suzie needs. This is a concern today, just as it was twenty-five years ago. Suzie's program is still severely under-funded, and LACL is faithful to find ways to bridge the gaps so that Suzie's supports do not suffer.

All of the above definitely played a role in Suzie's success; however, the role of Suzie's direct staff team cannot be overlooked. The ladies who care for her daily are the ones who have direct interaction through good times and bad. They make the "everyday" happen by faithfully coming to work and encouraging Suzie. They are the ones who developed the strategies to work through her behaviours and know how to comfort her after a night of seizures or when the frustrations of daily life overwhelm her.

It would not have been possible for any one person to achieve the results we see today, but as a team we make it happen. Naturally, over the course of the years the staff members have changed. Some stay longer than others, and some only left because they retired or moved away. Some years we had two or three staff members rotating through twenty-four-hour periods, and some years we have seen as many as eight members on the team.

We realized that it is not good to have any one person working with Suzie full-time, since it can cause burnout. It works best when no one works more than three shifts per week, with duties and activities divided among staff members as evenly as possible. Some of the staff can handle more days and activities than others, but we recognize the value of each person and what they bring to the team. We also realize that the value of consistency is so great to Suzanna that sometimes it is better to have a staff member who may not be so focused on achievement but instead is able to provide a sense of security for Suzie and has a good relationship with her as well.

We have not been able to keep relief staff around for very long. It seems very difficult to develop a lasting relationship with Suzie while working on an irregular basis. Either the behaviours will spike and the relief staff member cannot take it and moves on, or they get oriented at other programs where they are offered "easier" shifts. Some cannot take the wholehearted affection Suzie bestows on her staff members. Hugs and kisses are ever-present but may sometimes be outside of the person's comfort zone.

Suzie's parents are always involved in the hiring process. If the interview is for the supervisory position, they will be present at the interview. When new relief staff are hired, every effort is made for Suzanna's mom and/or dad to have the opportunity to meet the new staff and have some interaction with them before a permanent position is offered. Their opinion is always respected in selecting staff team members.

It appears to work best when we have more regular staff members working fewer shifts. We can then replace each other while on vacation or sick. It seems that occasional overtime is less costly than continuous relief staff orientations, which result in minimal staff retention. One out of every ten to fifteen staff members who are oriented stay, but those who do stay, stay for a long time and remain very committed to Suzie.

Since Suzanna is supported one on one and shift change time is very involved, with Suzie requiring full attention, the daily staff interactions are limited. The only way to ensure staff consistency is through regular staff meetings.

In the early years, the need for those meetings was so great that they were held weekly. Often there was no person available to look after Suzie. With her present during the meeting, all attention would have to be focused on her the entire time. In order for staff to be able to debrief with each other and talk about strategies, the meetings often took place late at night after Suzie was already in bed, or during weekends when Suzie was with her parents. Staff could then talk about how the week had gone, what worked and what did not. The ladies shared with each other if they discovered a strategy that worked for them, and all team members adopted it. If someone had a particularly difficult week, they could talk about it and receive team support.

Once LACL was able to secure the support of a behavioural consultant, the format of the meetings changed. Since we had an amazing consultant, Cynthia, we were able to breathe easier and be assured that the strategies adopted were OK. It was also good to have someone who was looking at events with a fresh eye, without emotions involved, and able to say, "Why don't you try this?" It often proved to be exactly what was needed for success!

Cynthia had weekly relaxation and counselling sessions with Suzie. That personal interaction alone allowed Cynthia to know exactly what we were dealing with and to use her professional expertise in a practical manner while setting realistic expectations. It also had one small downside -the sessions were very important to Suzie. After an incident, she would often say, "I HAVE TO keep it (anger) in me until I tell Cynthia." We did not realize this was happening until after Cynthia had to leave due to funding reductions, and Suzie started "letting go" of incidents faster. However, the benefit of sessions when we had them was much greater than any downside. Many valuable strategies were developed during that time. Cynthia also attended staff meetings, when necessary, to train and support staff. This was particularly valuable during times when incidents occurred more frequently. Cynthia recalls:

I first met Suzanna in my role as Behavioural Consultant in 2002, and immediately my professional boundaries softened. I was enamoured by her charm and enveloped by her warmth. On paper, she was described in strict behavioural terms: aggressive, agitated, threatening, but there was so much more to understand. The process of uncovering what triggered her upset and how best to support her when it occurred was a journey that her entire network went on together. The dedication of her family, staff team, and friends was unmatched. We met, discussed, debriefed, collected data, wrote plans, and made revisions. Certain strategies would be effective for a period of time, and others not at all. What we kept coming back to was the most basic of interventions: kindness and compassion. Suzanna at her core exemplifies these traits. Every meeting began with "Cynthia, you just won't believe it" and ended with a hug, a kiss for each cheek, and a candy for my pocket. Her expression of gratitude still overwhelms me, her heart immense. Despite the impact of her medical, psychological, and behavioural challenges, she continues to thrive and lead a meaningful life. She exhibits the power of resilience and reminds us all that when life becomes overwhelming, we can just "take a deep breath and try again.""

'Twas The Night Before Christmas

BY CYNTHIA CLARK

'Twas the night before Christmas,
When all through the Bailey home,
Not a creature was stirring, not even Sylvia Doane.
The family was gathering in the new house with care,
In hopes of some Holiday Tidings to share.

Suzanna and Sam were all snug in their beds,
While Sarah and Chewy stayed in Hong Kong instead;
And Mom planning menus full of turkey and sweets,
Dad, taking a nap, he's had too much to eat.

When out on the lawn there arose such a clutter,
They all ran to the window to see what was the matter.
Mom pulled back the curtains, Suzie opened the door,
And what they saw next made them fall to the floor.

In a miniature sleigh pulled by tiny reindeer,
That little old driver, St. Nick, did appear.
They danced through the air with a swoosh and a swish,
Just in time to deliver this one Christmas wish.

Suzanna, Samantha, Mom, Dad, and the rest,
If you run into problems, this idea is the best.
Just breathe and relax, count to ten if you need to,
Take a walk, have a nap, or ask for a "redo."

So to keep Christmas happy so you all have a ball,
Let your worries just dash away, dash away all!

Suzie's mom has always been involved in staff meetings, either attending them in person or consulting with the supervisor prior to the meetings in order to be aware of the agenda.

Suzie's family and Suzie herself are always actively and practically involved in the planning process. Suzie is aware of the goals that are set, and she often chooses goals herself. We will only proceed with a goal if Suzie understands it and chooses to pursue it.

Over the years, with the use of e-mail, text messaging, and ShareVision, the meetings' frequency has slowly been reduced. Currently the meetings are held every four to six weeks for staff and every three months for Suzie. If there are issues Suzie wishes to discuss with us, we will set up a "Suzie meeting." The agenda is put together by the staff team and Suzie, and Suzie chairs her own meeting. There are usually three to four items on the agenda in order not to overwhelm Suzie. She truly enjoys running her own meetings, and she handles them very well.

Suzie is also most accommodating during regular staff meetings. She will usually go to her friend Sylvia's house for dinner. However, before she leaves, she will often pull out mugs for us and start the coffee. She sure knows how to look after her favourite staff ladies!

DON'T COME BACK!

BY SYLVIA DOANE

Suzie will often phone me if she is upset or mad at someone, usually a support person. One day, she phoned screaming at a new staff person. Suzie was mad at her for some reason and had been chasing her around the house, trying to hit/kick her. The staff member finally ran out the front door. Suzie locked the front door and said, "Don't come back!" As Suzie was talking to me, she spotted the staff person's purse on the table and put the phone down for a minute. I could hear her laughing: "Ha-ha-ha-ha!" Suzie had thrown the purse out the door after her.

So what is it like to be a part of Suzie's team?

Despite all the consequences of medical interventions, including behavioural issues, left- side paralysis, and reduced abstract thinking, Suzie has not lost her zest for life. Her endless energy and a quickly discovered desire to help others, without fail captures the hearts of those who come into contact with her.

Suzanna's loving heart was, is, and always will be the one unchanging factor that draws people to her. It is no wonder that many members of her direct support team remain with her for many years, often staying closely involved even after their life path takes them elsewhere. It has often been noted by medical professionals that the continuity, commitment, and hard work of her staff members played the key role in Suzie's success.

Being with Suzanna for almost nine years now, I have to say that it is a reciprocal relationship. Without a doubt I would not be the person I am today if it was not for Suzie. I help her, but she helps me a thousand times more. She continually forces me to re-evaluate my system of values. When I came to work with her, I considered myself a good, honest, Christian lady. Since then I have often caught myself on hypocrisy, double-standards, poor work ethics, judgemental and unfair views of others, white lies, trying to take the easy way out, and many other things that showed me a "not so rosy" picture of myself. Did it discourage me? Maybe at times it did, but overall, it inspired me to change and become a better person. Suzie never gives up. She always tries and tries again, and so will I.

It is very hard, if not impossible, to remain fully professional when supporting Suzie. Heartfelt hugs and "I love you to death" statements are a part of everyday interactions. This is just who Suzanna is and how she expresses herself. You either accept it and reciprocate or very quickly find out that you are not able to work with Suzie and move on. Her heart is so full of love and caring that it simply overflows and affects those around her.

I would like to introduce some members of Suzie's team to you and share their stories, memories, and reflections.

Jane has been a part of Suzie's team from the beginning. Sunny Hill Health Centre for Children, Dainty House, and Benz Crescent are a part of her story. Jane has been Suzie's teacher. When the hemispherectomy robbed Suzie of everything she had previously learned, Jane was the one who, with faithful commitment re-taught her to read and write, complete math sheets, and many other skills. Jane has proven the doctors wrong when they said, "Suzie has now reached a plateau." She has done the "impossible." After her surgery, Suzie was not supposed to be able to read, write, or sing, and now she does it all! Although Jane has now been retired for a number of years, together with her husband, Dick, she remains involved in Suzie's life by attending her parties, coming to her Special O bowling and bocce nights, and participating in writing this book.

Sylvia has been a part of Suzie's team since the beginning and remembers the Sunny Hill, Dainty House, and Benz Crescent days. Sylvia quickly became a great source of support to the Bailey family, and I believe after more than twenty-five years is considered a family member. She has worked in the role of support staff, program supervisor, retired, and then came back as a staff member and retired again. Above all, Sylvia is Suzie's friend. Suzie calls Sylvia every night at 8:00 p.m. to tell her how her day has gone and to wish her a good night. (Suzie never missed this nightly call in over twenty years.) Sylvia has Suzie over for dinner every week. She is always there for Suzie. When Suzie is upset or frustrated, she calls Sylvia, and she knows her phone number by heart. Sylvia's husband, Wayne, and son, Wade, are also very close to Suzie and have often "rescued" us in times of staffing crisis. They help with gardening, installing the air conditioner in the summer season, and taking care of Lucy, the cat. For several years they tirelessly collected bottles for Suzie's Christmas project in the apartment in which they live. Suzie reciprocates by looking after their cats and home when they are away. Sylvia has always been a "decorating queen." She makes sure Suzie's house is ready for the holidays, marking Thanksgiving, Christmas, Easter, St. Patrick's Day, and Valentine's Day. You can definitely follow the rhythm of the year at Suzie's house, thanks to Sylvia!

Jody has worked with Suzie for over twenty years. Jody works with Suzie three mornings per week. She has raised her family during those same years. Suzie impacted and helped shape Cassidy and Benjamin into the lovely young people they have become. Jody's family attends Suzie's Christmas open house and her birthday barbecue. Jody's husband, Brad, helps out by putting up Christmas lights and manning the barbecue. Jody is a very loyal member of the team and prides herself on her work ethic, always arriving early for her shift and often staying longer to ease the shift transition. She never misses work, except for pre-scheduled vacations, and attends staff meetings regularly to be on board with every strategy we may undertake. She is a steady, reliable force in Suzie's life and assists with the work of the team, training new staff, supporting Suzie's community involvement and various employment/volunteer opportunities over the years.

Ami worked with Suzie for close to fifteen years. Ami balances two jobs. She works in school during the week and then on the overnights with Suzie. Ami creates a relaxing atmosphere and knows how to balance Suzie's busy life with times of relaxation and comfort. Ami's family loves to host Suzie for a holiday or birthday dinner and spoil her with lovely gifts at those special times. Ami also adds a "grain of common sense" to our team, helping adjust expectations to a more realistic level.

Vonna worked with Suzie full-time for over two years and moved on to become a program supervisor at another site. Vonna persevered through a lot of initial difficulties with Suzie and came out victorious at the other end. This led to many wonderful times and a strong relationship, which is still lasting even though Vonna no longer works with Suzie on a regular basis. Vonna makes a constant effort to remain involved with Suzie, comes for an occasional overtime shift, and visits with Suzie at her new program whenever possible. Vonna brings a special touch to those visits by bringing her dog, Talia, whom Suzie adores. It was Vonna who helped Suzie get started on the Christmas Bureau project, leading Suzie to fame and the front pages of *Langley Times, Langley Advance,* and the *Vancouver Sun*. Vonna still faithfully collects cans and bottles to donate to Suzie's project.

Bridget is another lady who holds two jobs and has remained faithful in her commitment to Suzie over the past nine years. Bridget is very professional and often adds valuable input during staff meetings and planning sessions. She has a rare ability to "bridge the gap." (Is it a coincidence that her name is Bridget?) When team members have opposing ideas, Bridget can be counted on to help find common ground and get the focus back to what is important.

Eliza has been working with Suzie for almost five years. Eliza was Suzie's neighbour in the Woodbridge complex before coming to work for LACL. When her first orientation shifts were scheduled here, she knew it was meant to be. Knowing how critical it is for Suzie to have as consistent a staff team as possible, she persevered through some rough times in the beginning. Eliza's husband and children also quickly learned to love Suzie and opened up their lives to her, coming to Suzie's parties and visiting with her at other times as well. Eliza's husband, Darmon, generously invited Suzie to set up her new business (a vending candy machine) in his company's office. He and his office staff are very committed to Suzie's success and cater to her needs at all times, adjusting the work of the office when she comes to make sure each visit is a success.

Regina joined the team over three years ago. She is well-loved by Suzie and the team. Suzie bonded with Regina almost without any of her usual testing time and treasured every shift they had together. Regina brought a lot of warmth and caring to the team. She has the gift of encouragement that is very special to all of us, especially when times are hard. Although she has moved on to a supervisory position in another LACL program, she remains in touch with Suzie and the team and comes to visit as her busy schedule allows.

Sara is one of the newer members of the team. She is a quiet girl with a very big heart and excellent work ethic. Suzie quickly bonded with Sara and sincerely enjoys their time together on weekends.

Michelle is the newest member of the team. She has quickly become a part of our team with a big commitment to Suzie and excellence in her supports. Michelle is a tiny young lady with great courage and a big heart. She braved the initial difficulties and is comfortable supporting Suzie even in high trigger activities, as well as taking her to musical concerts.

Nightly Phone Calls

BY SYLVIA DOANE

When Suzie calls me, she has the need to start the conversation off by saying, "Sylvia, you stupid asshole," or words to that effect. I tried a few different ways to change her greeting manners, for example asking, "Can you remember to talk to me nicely when you call me?" —all without much lasting success. Finally, I seemed to have found a strategy that works. I have a call display on my phone, so as soon as I see that it is Suzie calling, I answer the telephone by saying something like: "I sure hope this is my friend Suzie Bailey calling to say hi to me and see how I'm doing?" She always answers: "Yes, it is," and I respond: "Thank goodness! Nobody has called me all day except you, Suzie. You must really be my best friend!" Her reply is always something positive, like "Of course I am, you cutie patootie." The conversation moves on in a much happier tone with this kind of positive beginning!

Here are some thoughts, memories, and reflections we would like to share about what it is like to work with Suzie and how she affects our lives:

Did you hear of Suzie and what working with her would entail before you were interviewed? How did you come to work with Suzie? Had you met her before

And, if so, in what way?

- Jody: I found an ad in the newspaper. I was looking for a job in the field after I had my kids. I worked in the field before for a group of parents who took their kids out of Woodlands as young adults. I worked there for about five years before I had my kids. I was a supervisor there, but I did not want to supervise anymore. The ad in the paper said it was to support a young lady with severe behavioural challenges, and that's exactly what I was looking for. Initially, I worked Saturday evenings, Sunday mornings, and Monday evenings at Benz Crescent.
- Bridget: I didn't know anything about the program. I interviewed as casual and was directed to high behaviour programs. I came to orientate with Suzie and stayed there as relief, eventually taking on a small part-time position.
- Eliza: I actually met Suzie even before I applied with LACL. She used to be my neighbour in the townhouse complex where she lives now. I would walk my dogs, and she would be out cleaning the complex. We would chat on occasion. Even beforehand, I met her at a White Spot restaurant in Aldergrove, where she often dined with her mom, and I would serve them.
- Michelle: I applied for a relief position with LACL and was offered a position with Suzie soon after my interview. I didn't learn any details about her program until I met with the supervisor for my initial orientation.

What were your thoughts after the first several shifts? Were you surprised in any way?

- Jody: I was more impressed that Suzie was so helpful and that she was very gracious in letting me into her home. She opened her life to me, and I respected her for that. I wanted her to know that I was there for the right reasons. She's had a lot of people come through her life. At that time, they had trouble staffing the house, so I was expecting the worst. But I didn't get the worst. Suzie is very forgiving of our mistakes. She knows that we are trying. You make yourself vulnerable because you don't know everything. You ask Suzie for help.
- Ami: It was very hard. I wasn't sure I was going to make it. It was a totally different experience. I came from the Autism program. The expectations here were totally different. Other staff and the program supervisor were all very helpful and supportive of me.
- Bridget: I was inspired by Suzie's ability and independence with the amount of disability she had. I thought she was very kind and loved the fact she was so independent.
- Regina: I wasn't surprised, but I thought in my head, *This is not going to be easy. It will be tough.* But the way Jody portrayed Suzie's qualities during our orientations convinced me that this was my place. She said a lot of things about Suzie in a strong, positive way.
- Sara: I was sort of surprised. I was bracing myself for the worst. I was expecting her to test me more. It went better than I anticipated. Even up until now, we didn't have any bad days, only moments. But she got over them quickly. She surprised me that way.
- Eliza: My thoughts after the first several shifts were that I was really blown away with how compassionate she is, because she is so willing to do things for others. She finds joy in it. She has a selfless nature to her. I was surprised how challenging this line of work is, supporting somebody. It does challenge you as a person to make the shift successful, because you need to think four or five steps ahead.

- Michelle: I was surprised by how well my first few shifts went with Suzie. I was a little scared after my initial introduction with the supervisor, because I was told some of the things that could potentially go wrong, so that I was prepared and knew what to expect. I had never had to deal with that kind of thing before, because I had started in the field dealing with disabilities more on the physical and medical side of things. But Suzie was very welcoming and patient with me and made me feel comfortable in her home.

What was different about working with Suzie versus other one-on-one programs (e.g. the home modeled to meet Suzie's needs as opposed to Suzie having to fit into a typical group home model)?

- Jody: It was more residential, more life skills based, and personal care. My previous experience was a day program. It was a lot of firsts for someone who has never experienced things that we consider normal, like going out for a cup of coffee and grocery shopping.
- Bridget: I liked that Suzie was able to advocate for herself. She has a say in what happens in her life. All staff respect Suzie. I liked that the staff gave her time to respond and make her own decisions and the space to do activities for her. Suzie will come out and tell people what she wants and needs, and the staff are kept accountable to her.
- Eliza: The standard of care at Suzie's house was high. The team really came together to make her life successful. The team that works there is absolutely unbelievable. I knew that it was what I wanted to be a part of. I wanted to help continue making a difference in her life.
- Michelle: Suzie's team is amazing and it makes all the difference in her life. You can tell from the second you step inside her home that everyone there really cares about her and that she cares about them. Every day is structured around what's best for Suzie instead of what's easiest or most convenient for everybody else.

What do you know of her early history and her seizure challenges, surgeries, and medical issues?

- Jody: We were at Children's Hospital for a spinal tap procedure. They were doing an epidural on her and she was so uncomfortable. It was very painful. I sat close to her and held her hands through it and I was just talking to her. The nurses said, "You have a relationship with that girl; you're not just a worker." Suzie won everyone over by that. At one point she had a seven-week stay at UBC Hospital and we demonstrated the same. All of us staff ladies had a relationship with her. It was the same after her last brain surgery. We were waiting for her in the recovery room. When she was coming out I remember one of the medical personnel saying, "It was a difficult surgery." This was because of all the scar tissue from previous surgeries. I felt a presence there in this room and I knew she was going to be OK.
- Eliza: When I first started there, I had the opportunity to read an article called: "Oh Suzanna! A Nursing Challenge" written by two nurses. In this article it described Suzie's childhood as a little girl with her seizures and what she went through with her surgeries. It was remarkable the things that she went through and how she overcame a surgery that would change her life. The surgery was to prevent the seizures from occurring. However, it resulted in an acquired brain injury. I knew this lady had great inner strength. I knew I could learn from her.
- Michelle: I was given some details about Suzie's childhood and medical past in my orientations. I remember being really sad for a while, thinking about all the things she's been through and all the challenges her medical history has presented throughout her life, but then realizing how far she's come since those earlier days and what an amazing accomplishment it's been for her to build such an active, fulfilling life for herself despite all of it. It's really quite inspirational.

What influenced your decision to stay?

- Jody: I never thought of not showing up. I didn't go through some of those super hard times and experiences that other people had.
- Ami: Once I had built my relationship with Suzie (which took a couple of months), I got to know her more. Before that I had a really big incident with Suzie and was off for a time. Suzie felt really bad about it and when I got back it was easier. Suzie treated me out for a dinner with Sylvia. I got another orientation then and from that time Suzie was much better with me. I could handle my shifts. I was very confused and didn't know what I was doing, but I had a mortgage. I needed the job. Without the support of my friends at work I could not have survived. At some point two of her regular staff left and that increased Suzie's trust in me. Our relationship greatly improved from that point on.
- Michelle: The longer I stay with Suzie the more I realize how difficult it will be to ever leave her. When I first started I was told that once Suzie started to trust me, things would get easier. What I didn't realize is how much we would come to trust each other and start to build a mutual bond. It's impossible to have a one-sided relationship with Suzie. Suzie truly loves her staff, and we all love her in return. I've learned so much from her that I never expected to.
- Eliza: My decision to work with Suzie was influenced by the wonderful team that works there and their compassion. Suzie humbles me and teaches me how to be grateful within my own life. Every shift she touches my heart or makes me laugh.

What was or is your biggest challenge and what was or is the hardest part of supporting Suzie?

- Jody: I think just going into environments out in the community. I used to take her to rhythmic gymnastics. She still struggles with that today, but in those days her "entrances" were something else.

- Bridget: I think the biggest challenge in supporting Suzie is to stay one step ahead of her in order for her to be successful at any given activity. When Suzie isn't successful we can often look back at the situation and see where we could have done things differently in order to make the activity successful. For example, I took Suzie to music therapy once, which would alternate week by week with karaoke. I had never taken her before, but was prepped and assured it would be fine. Suzie had said it was karaoke day, so I prepped Suzie this way, all morning saying how excited I was to see her sing, but when we arrived, it turned out it was music therapy. Well, Suzie lost it. She could not handle the unexpected change. She blew up at me, and I could not calm her down. I tried to direct Suzie to leave but could not get her to leave. After some time, I was able to get Suzie in the back room. We tried to call another staff member for her to talk to, but no one was available and that just escalated her more. Finally, a staff member came into the building and to our aid. It took quite some time to get Suzie calm and reasonable. I felt really bad for Suzie, not because she was upset with me, but because I felt bad for her disappointment. She was so eager to show me how she sings. I could have totally prevented the incident from happening by just calling ahead to the program to confirm if it was karaoke or music therapy. I was near tears the whole time, feeling so, so bad that I had put Suzie through that. She then apologized to me for her actions when I was the one who should have apologized, which I did too.
- Regina: I'm so lucky. I have had only one incident with Suzie, but it is about being prepared for what could happen. You have to try to foresee what is coming. You have to plan on how you will respond. If she is escalating, I listen to her and give her space. I don't talk when she is talking. I don't interrupt. I think most of the struggles come from bad communication. She wants to be in charge, and you need to give her an opportunity to do that and to

voice everything that's on her heart. When we listen to her, she knows we are there.

▸ Eliza: When I initially started working with Suzie, I was really eager to do a lot of things in the community with her, whether it was interacting with friends or an activity she enjoyed. I remember being very eager to do these things, and I almost went too fast because I wanted to do so much with her. I was still learning and understanding Suzie.

The very first incident I had with Suzie was in PetSmart. Before I went to the pet store, we made a trip to the pharmacy to drop off a blister pack that contained medication that needed to be returned. We went in there, and Suzie handed in the blister pack as I waited outside, because this was one of the things she does independently. I didn't realize the significance of her perception of the blister pack. The blister pack wasn't finished. In her world, she needed to finish it. She couldn't understand that it was a medicine change, and we were getting a new blister pack. This triggered her. I recall her becoming upset with Margaret from Valley Pharmacy, so I went inside to help Suzie out when she was in distress over this. I explained to Suzie that she would be getting a new medication and helped her to take ten deep breaths. She apologized to Margaret and we were able to leave the pharmacy.

When we were in the car, I asked Suzie if she would still like to go to the pet store or if she would like to go home. Suzie said she still wanted to go to the pet store. On the drive to the pet store Suzie was silent, which was unlike her. At the time I thought I was successful, but when I reflected on it later, I realized that she usually sang along to the CD in the car, which she did not do on that day. Once we were in the pet store, Suzie helped me pick some dog bones. (We ask Suzie to help us out to reinforce and develop a positive relationship with her.) We were standing at the counter getting ready to pay and I remember that Suzie started to swear.

She was swearing because the conveyor belt was moving too fast, which is one of Suzie's triggers.

At the time I didn't realize this was one of her triggers, and I remember trying to correct her language in the store because there were children behind us. Suzie did not like me reminding her about her language. She reacted by grabbing me by the back of my head and slamming my head against the conveyor belt. Then I panicked. I paid the lady fast. I was thinking I could re-direct Suzie and get her into the car. As I started walking out of the store, Suzie was very agitated with me and I became her target. I was walking backwards with Suzie following me, telling her it would be okay. I still remained close because I was mindful of all the safety issues that could happen. Suzie again got a hold of the side of my head, because I was too close. She was pulling at the hair on the side of my head. She was kicking, spitting, and swearing at me out of frustration.

A stranger came and saw this happen and tried to help. As soon as the stranger tried to intervene, Suzie directed her anger towards her. It happened so fast, the stranger brought Suzie down. While all of this was happening, this lady's girlfriend called the police. Then I got on the phone and called senior staff for back-up. The police came and talked to Suzie and no charges were laid. I went to the car to get Ativan for Suzie and came back to give it to her.

The senior staff had come by that time and drove Suzie back to her house. I remember thinking: *Oh, my goodness! I don't know if I'll be able to do this.* But with Suzie's team being so supportive of Suzie and each other, we were able to repair that relationship, even though Suzie was so upset with me. Suzie sent an apology card to the police officer and the lady who intervened at the pet store. Currently the hardest challenge in dealing with Suzie is how quickly she can react and how fast things can change. Also, I am quite a bit smaller than Suzie, which makes me have a little bit of a guard up.

- Michelle: My hardest incident with Suzie occurred while we were doing her weekly job of cleaning her complex. She loves to help her neighbours pick up all the garbage that's on the ground and make the neighbourhood look beautiful for them. It's such a selfless motivation that inspires her to do this, and she's usually outside rain or shine. On this particular day there were two neighborhood kids playing with Nerf guns outside. Suzie noticed one of their darts on the ground and wanted to put it in her garbage cart. Before I could react, one of the boys noticed Suzie pick up and throw away his toy and ran over to take it back out of the garbage. Suzie swung her garbage picker at him and hit him in the head. He started crying and crouched on the ground, holding his head. He wouldn't even walk near her after the incident and ran home the long way. It was so hard for me to remain calm, and I was near tears calling the supervisor. In my world, it's never OK to hit a child, and I felt so terrible for him and couldn't believe she did that, but Suzie was just trying to do her job and didn't understand why the boy would try to stop her from doing that. So I think the hardest part of working with Suzie is controlling my own emotions in an escalated situation and understanding that she doesn't always process the world in the same way and isn't always able to control herself when she's upset, which can cause her to do things that I in turn don't understand, and it's not her fault. To her immense credit, she has since written that boy a card and brought it over to his house to apologize. Those boys now always greet Suzie when they see her outside, and she saves their Nerf darts in her bedroom drawer for them whenever she finds them outside.

What is your favourite part? What do you like most about Suzie?

- Jody: Oh boy, that's a big one. How can I put that into words? I love being with her. Whatever our day looks like, I genuinely love coming in. I just love, especially when I'm with her for three days, to give her everything I've got. We are a team when we are together. I've been told in the community that we're like an old married couple. It means going through the good, bad, and ugly and still respecting each other on the other side. She has made me a better person; she's made me humble. She makes us get out of our comfort zone.

- Bridget: Suzie's genuine compassion and caring for others. She is always thinking about someone else and how she can help them. One time I was driving with Suzie to do the mail run for head office. It was the time when my husband wasn't well with his heart condition. My phone rang while we were driving and I asked Suzie if I could pull over and answer it in case it was my husband. Suzie did not hesitate and said, "Of course, Bridget, you can pick up your phone. I can help your husband out." It turned out it was my son, and my husband was headed up to the hospital and I had to leave work to go there. Suzie was so quiet and calm in the car as I called the supervisor and explained the situation. Then when I explained to Suzie what was happening, she said, "Don't worry, Bridget, I can take care of your husband. I will keep him in my prayers. Don't worry about the mail run. I can do it with the other staff, no problem." She kept repeating how she was going to take care of my family and me! These were such comforting words at that time for me. It still brings me to tears when I talk about it. Also, I love her sense of humor. You can always count on Suzie to have a good laugh!
- Regina: Every shift is different; you don't know what to expect. She can be so funny. You never know what is going to happen. That surprise element is great. But I always come in with a positive attitude.
- Sara: I just find her so inspiring. She goes through so many challenges, struggling with seizures, but she manages to remain positive. She enjoys listening to her music, and dances for me. She overcomes obstacles and overall always focuses on the positive in the long run. This may not be so visible in the short term, but

in the long run she does. My favourite story with Suzie so far is this: For the first several shifts, I didn't know I was supposed to fill the water cup for her routines. Suzie never mentioned it, but after seven or eight shifts she came out of her bedroom and said, "Sara, I just wanted to tell you this. I don't want to get mixed up about it, but I want to let you know you always forget to fill my water cup. You should better remember for the next time so that we don't get in a fight."

- Eliza: I would have to say the thing I like most about Suzie is her heart and determination. She was given a raw deal but never lets it slow her down. It's really remarkable all the things she has accomplished. Watching her change and touch people's lives, so selfless and kind, is something we can all learn from.
- Michelle: Suzie is the most selfless, caring person that I think I've ever met. Almost everything she does is to help somebody else out. She doesn't even differentiate between people she sees all the time and loves or strangers that she sees on the road or in the community. She cares about everybody, and I learn so much from her about that.

What do you think played the biggest role in Suzie's success?

- Jody: Team effort, without a doubt. Everyone brought something different to the table, how we looked at things. We were not afraid to pass this on.
- Bridget: I've known her for the last eight years, but from what I have seen, I would say her close family connections. They advocated for her so well, for her to maintain her independence. Also her dedicated, long-term support staff, the people who truly love and support her and all her positive contributions to the community life!
- Ami: Everyone's involvement, the approach of all the staff in her life and her family. She is a very giving person. I have learned so much from her. She taught me patience and how to be a good

listener. I used to go to church, but I didn't have much faith. My faith grew and got so much stronger and deeper because of Suzie and the way she prays. Even my kids saw Suzie had really good qualities, such as the way she gives of her time. She talked to them on the phone and had a really big influence on them.

- Regina: I'd say people around her: her parents, staff, and friends. When I see what Sylvia is still doing for Suzie after all those years, it motivates me. People going beyond the call of duty are what keeps Suzie going—Jody, Ami, Bridget, Eliza. How do they do it? I'm not sure if I would continue coming if I had to go through what some of the ladies went through. But their commitment is what defined Suzie's success.
- Eliza: I would have to say her family and staff have all come together for this angel on earth. And I have said from day one, Suzie's staff members are one of a kind. And it's a pleasure to be a part of her team. It brings me a lot of joy. The successes in her life outweigh anything else. That's why I'm working with her today. With Suzie's determination and heart and some good people at her side, anything is possible. Just read her story!
- Michelle: All of the people involved in Suzie's life are amazing resources. We all care so much about Suzie, and her best interests are always first and foremost in all decision-making. Her own determination and the way she speaks up for herself and makes her wants and needs known are also huge factors.

If you could change one thing in the way we support Suzie, what would it be?

- Bridget: We could always work more on communication among staff and being on the same page to prevent incidents and avoid situations that could be triggers. But really, I wish we could do something more to help get her seizures under control and help her extreme fatigue so that Suzie can be Suzie and do all the really great things for others that she so enjoys doing!

- Regina: Suzie has been through a lot, but I think about Suzie's language. If she is rude with me, I will always go back and talk to her about it. The words that she uses sometimes stay with us, and we say them too. I wonder where this comes from. I know we let Suzie say them in the privacy of her house, but I think we shouldn't. We shouldn't encourage them at all.

Self-Check-Out

BY AGA KARST

Suzie and I stopped at Save-On-Foods to pick up a few items that were missed during the weekly grocery shopping. Since we only needed a few items, I suggested we use a self-serve check-out. A blank look on Suzie's face triggered my question: "Have you used the self-serve check-out before, Suzie?" "No, I haven't," came a quick reply. "Would you like me to teach you how?" "Yes, please." We approached the register. I instructed Suzie to press "Start" and quickly located the More Rewards card in my wallet. In the meantime, the pre-recorded voice in the register announced, "Please scan your More Rewards card." Suzie looked confused and yelled, "Shut up, you lady!" I did my best explaining to Suzie that there was no real person involved, only a prerecording like on her CDs. I could tell that Suzie had a difficult time grasping the concept. She managed to contain her irritation and followed my instructions to complete the transaction. When the recording announced, "Thank you for shopping at Save-On-Foods," Suzie, with a big smile replied, "You are very welcome, you gorgeous lady! Thank you very much for your help!" I wanted to give Suzie a big kiss and a hug, but all I could afford in the public setting was a smile. "Lovely manners, Suzie!"

YOU JUST WON'T BELIEVE IT!
BY BETTYANNE BATT

Today, Suzanna Diane Monique Bailey (Suzie) lives a wonderfully full, active, and rewarding lifestyle, nurtured by family, friends, and a team of dedicated staff. To look back, however, one would wonder how this could be possible given the multitude of medical issues and resultant behavioural challenges that have confronted Suzie from early childhood to the present.

We have met as a team twice a month over several years to document the amazing challenges and successes of Suzie, her family, and those who care for her as staff and close friends. What a journey this has been!

We came to understand the years of worry, fear, hope, and despair that were a constant in her parents' lives. We heard of their concern for Suzie's sisters as well. We were also told that their deep spiritual support came from their faith, family, and friends, and that this was of invaluable assistance in those dark times. Over and over again, they were enveloped with professional care from their medical teams and support from dedicated, caring staff, all of whom contributed to providing an environment wherein Suzie could flourish.

Were they shocked at the physical, intellectual, and psychological damage resulting from her six neurosurgeries and the countless side effects of medication? Were they shocked at the behaviours that became part of Suzie's everyday experiences? Absolutely! But their unconditional love has overcome this. Suzanna has a strength of character that has beaten the odds over and over again. She is able to give back to her community, and she does this so well. Her contributions are valued by many. Caring for others is her signature on life.

It was always our intention as we met, talked, and jotted down our thoughts that our readers would see that what seemed like impossible issues to resolve were overcome. There were no easy answers to guide us along the way. We learned that we must keep trying, often in many different ways. Of great importance was being strategic, proactive and constantly providing Suzie with teaching cues that would assist her in enjoying positive social experiences. We also learned that many issues are lifelong and would need continual support to ensure daily challenges are addressed successfully. There was no quick fix, but perseverance and consistency in teaching experiences, offered in a gentle and caring home environment, would lead to successful pathways for change.

Those of us who are close to Suzie and her family often wonder what it is that allows Suzie to be so accepted and successful. What is it that helps her beat the odds of everyday living? What is the ingredient that provides the tremendously welcoming impact she enjoys within her community, despite the challenges she lives with day in and day out? The answer? It is Suzie herself! She has a heart of gold and is motivated simply by her wish to help others. Suzie sees the good in everyone she meets and knows no prejudices. What comes so honestly from Suzie is her inherent warmth and charm. These characteristics have no bounds and provide the means for her to be the helpful person she desires to be.

Suzie is nurtured by her family who cherish and love her. They are always present and involved. Setting the bar high, they support her to live in a milieu that provides the best opportunity for happiness and success. They have never hesitated to take a risk despite the outcome.

All of the above definitely play a role daily in overcoming Suzie's challenges, but it must be noted that underlying her successes has been effective communication that is as important today as it was when she was a child.

If, after reading this story, you are able to find a renewed sense of hope or a reinforcement or confirmation of your efforts, or find a strategy that may assist you in your future support efforts, then we have achieved our goal.

We wish to thank Suzie for granting us permission to write her story, along with support from her parents. We are grateful to all who have offered us their time and expertise, and to those who allowed us to include them in Suzie's story. We are especially appreciative of the stories from her staff, past and present.

Thank you!

ACKNOWLEDGEMENTS

The authors would like to express their thanks to the ERIN Research in Neuropsychiatry Fund that provides support to the academic programs of the British Columbia Neuropsychiatry Program and supported this project to improve the public awareness of and discussion about the body-mind connection.

This book would not have been possible without the contributions, support, and encouragements of many people.

*To the Bailey Family for their encouragement and support

*To Dr. Kevin Farrell and Dr. Trevor Hurwitz for their dedication to Suzie over these years, and for their contributions to this book

*To the Langley Association for Community Living for their support

*To our first editors, Cathy Bergvinson and Marian Mahony, for their assistance and professional advice in polishing our manuscript

*To all those who contributed in any way to the writing of this book

*To our families who allowed us the time we required, and finally

* To Suzie, who says "I'm STILL the Bob and Cathy's Kid!"

THANK YOU!

ABOUT THE BOOK TEAM

As you read our story, you will notice that we reference interviews and articles written by the "Book Team": Jane, Sylvia, Aga, and BettyAnne.

When the Bailey family requested assistance from our agency, the Langley Association for Community Living, it was decided that a home with a family-like type of environment was to be designed so that care and support for Suzanna and her parents could be offered in a holistic and caring manner.

The staff of her first home (Benz Crescent) were instrumental in these early years of Suzanna's life, as they attended to her needs with compassion and caring. Over the years, many staff members offered support and guidance for Suzanna, but two ladies (and their families) who have been most influential were Jane and Sylvia. Both were staff supervisors and have known Suzanna and her family since she was ten years old. Their long-term relationship with the Baileys provided valuable insight and recollections as groundwork for this book. Today, both Jane and Sylvia and their families still include Suzie in many social events.

BettyAnne was the Respite and Family Support Coordinator for LACL and offered administrative support until Suzie reached adulthood. BettyAnne continues to applaud Suzie's successes up to this day. She provided gentle direction to the whirlwind of ideas, often guiding the team to find a solution pleasing to all when differences of opinion occurred.

As Suzanna moved into her own townhouse complex, several key staff along with Sylvia continued to provide guidance to ensure Suzie was reaching her potential. Upon Sylvia's retirement, Aga became her staff supervisor, and an instant bond developed. Building on the groundwork set by Sylvia and Jane, and with an understanding of Suzie's challenges and strengths, she supports the staff to ensure that Suzie has a meaningful and full life.

TREVOR HURWITZ, MBCHB, MRCP (UK), FRCP (C)

Dr. Hurwitz completed his medical degree in Pretoria, South Africa followed by an internship in Johannesburg and a senior houseman-ship in Cape Town. He continued his studies in London, England and obtained certification in Internal Medicine. He completed a residency in Psychiatry at the University of British Columbia in 1980 and a residency in Neurology at Boston University in 1982.

Currently Dr. Hurwitz is a Clinical Professor in the Department of Psychiatry at the University of British Columbia. He is the Director of the B.C. Neuropsychiatry Program and Chief-of-Service of West-1 at UBC Hospital, Vancouver. Dr. Hurwitz has a joint appointment in the Department of Medicine, Division of Neurology and practices as aclinical neurologist in addition to his primary commitment to inpatient and outpatient Neuropsychiatry.

Dr. Hurwitz's work is mostly devoted to clinical practice, teaching and the promotion of the discipline of Neuropsychiatry with some research interests in the field of Neuropsychiatry.

Photo credits to Michelle Karst Photography

"The Book Team"
Left to Right:
Aga Karst
BettyAnne Batt
Jane Huff
Sylvia Doane
Dr. Trevor Hurwitz

ENDORSEMENTS

I wish this book had been available in the early 1980s when the Vancouver School Board became involved with the mainstreaming of special needs students and the subsequent closure of the facilities for these students. I was the principal of a receiving school and I would have been so much better prepared if I had read this book. Little did we know what incredible feats of dedication and support were necessary for the successful integration of these children? Indeed, the publication of this book is still timely and essential.

A. W. Paterson
Principal (Retired)

This is an inspiring story of a remarkable young woman, who in spite of having a severe seizure disorder and a brain injury, has triumphed and is living her best life. The book is very informative and real! As a parent of an adult son with autism, I felt the anguish that goes with parenting a child with special needs. My son shared a home with Suzie for three years. The lifelong relationships that the authors developed with Suzie are very special! I would highly recommend this book and I commend the family, friends, and those people closest to Suzie for the work they have done!

Sincerely, Judy Forster
Parent of a son with autism and former Mayor of White Rock, BC

"This book is an amazing effort, about an amazing person and family, and the amazing community around her that cares for her. Suzanna Bailey's story is one that shows the deep spirit of community, and how seemingly the most challenging of situations can be addressed through the compassion and adjustment of family, friends and systems. For people (and families) who experience disabilities the deep drive is to try to fix the problems that are brought on by disability – this is a natural "micro-based" response. The countless surgeries, medicines and treatments that Suzie went through in her early years are evidence of this drive. Yet for Suzie Bailey, the magic of her success in community today has been the "macro" adjustments that have been made by all who have come to know her. As people have developed relationships with Suzie, they have come to accept her just as she is – the good and the bad – and make accommodations that have led to her community success. This book is a must-read for any family or professional who experience or work in support of people with disabilities. The lessons in these pages are astounding!" Doctor Al Condeluci

Al Condeluci, PhD

Dr. Al Condeluci has been a leader in human services and disability advocacy work for the past 45 years. Holding a PhD and MSW from the University of Pittsburgh, Dr. Condeluci has been the CEO of CLASS (Community Living and Support Services) a major non-profit organization in Pittsburgh, PA since 1973. He holds faculty status at the University of Pittsburgh in the Schools of Social Work, and Health, Rehab Sciences. He is author of 7 books including the acclaimed, Interdependence: The Route to Community (1995) and more recently, Social Capital: The Key to Macro Change (2014). He serves on a number of nonprofit boards and government commissions on state, local and national levels. He helped found, and convenes the Interdependence Network, an international coalition of professionals, family members, and consumers interested in community engagement and macro change. He can be reached at **www.alcondeluci.com**, or @ acondeluci on Twitter.

APPENDIX A

Functional Hemispherectomy

One of the surgical procedures that Suzanna was subjected to is called functional hemispherectomy. It was developed as a result of various trials of anatomical hemispherectomy (initially performed in 1928) which led to the numerous complications.

"The first anatomical hemispherectomy (the removal of the cerebral cortex of one hemisphere en bloc or in segments) was performed by Dandy in 1928. Its original purpose was a dramatic attempt to cure patients suffering from malignant gliomas of the cerebral hemispheres. The procedure was abandoned in the late 1930s when neither cure nor prolongation of life was achieved. Dr. K.G. McKenzie, at the Toronto General Hospital, performed the first anatomical hemispherectomy for the treatment of seizures associated with infantile hemiplegia in 1938. Despite promising results it was not until 1950 when Krynauw published his impressive results on a series of patients treated by this procedure that it became generally used throughout neurosurgical centres. The first anatomical hemispherectomy at the Montreal Neurological Hospital (M.N.H.) was performed by Dr. Penfield and Dr. Rasmussen in 1952.

The procedure was abandoned when reports by Laine, Pruvet and Ossen in 1964 and by Oppenheimer and Griffiths in 1966 showed that the syndrome of superficial cerebral hemosiderosis was a late and often fatal complication of the anatomically complete hemispherectomy. Occurring at a mean time interval of eight years following their operation, patients presented with a progressive neurological deterioration associated with obstructive hydrocephalus and increased intracranial pressure due to toxic effects of iron on the brain. Pathological characteristics included:

1. A membrane lining the hemispherectomy cavity and the ventricular system similar to those seen in chronic subdurals.
2. Brownish cerebral spinal fluid with an increased protein and iron content.
3. Iron pigments in the cortex, basal cisterns and subarachnoid spaces. (2)

It was postulated that the large cavity remaining post hemispherectomy did not provide adequate physical support to the remaining brain. Repeated small traumas as a result of such things as sneezing, coughing, and sudden head movements resulted in hemorrhage and seepage of red blood cells into the cavity and ventricular system resulting in the above syndrome. To eliminate this serious late complication, the anatomical hemispherectomy was replaced by the subtotal hemispherectomy." (Excerpt from "The History of Functional Hemispherectomy" by B.Taugher and M.Richards, March 1991)

Due to these undesired side effects, Functional Hemispherectomy for seizure control was developed by Dr. Rasmussen. Instead of removing the brain tissue from the right side of the brain completely, the procedure involves a disconnection of that hemisphere, while leaving the brain tissue intact.

"This surgery was based on the concept of partial anatomical resection and complete physiological disconnection of the residual unresected brain. The long term complications of AH were avoided to a large extent with this procedure and provided comparable seizure results." (Dr. Roy Thomas Daniel, "Hemispherectomy")
https://www.epi.ch/_files/Artikel_Epileptologie/Daniel_2_03.pdf

This was the type of surgery that was available during the time Suzanna needed. The procedure has since been developed and improved. For readers interested in further, more current reading on the subject are encouraged to check the following links for more detailed information:

Hemispherectomy: Indications, Surgical Techniques and Outcomes
https://www.omicsonline.org/open-access/hemispherectomy-indications-surgical-techniques-complications-andoutcome-2155-9562-1000300.php?aid=59363

Peri-insular hemispherectomy in paediatric epilepsy by Jean-Guy Villemure
https://www.ncbi.nlm.nih.gov/pubmed/8559348

Neurol Med Chir (Tokyo)
https://www.jstage.jst.go.jp/article/nmc/46/4/46_4_182/_pdf

APPENDIX B

Dr. T.A. Hurwitz 2001 Report

DR. TREVOR A. HURWITZ, INC.
MBChB, MRCP (UK), FRCP9C).
Neurology and Psychiatry
27 November 2001
Re: Suzanna BAILEY

Suzanna Bailey is a 20 year old right-handed female last seen by me in August 1997. She lives in purpose built group home where she receives 24 hour 1:1 care under the aegis of the Langley Association for Community Living and funded by the Ministry of Social Services and Housing. She has an intractable seizure disorder treated by functional hemispherectomy. In May 1997 she underwent her 5th neurosurgical procedure removing the last remaining right hemisphere tissue.

Her seizure disorder has been complicated by a disinhibited frontal lobe syndrome characterized by interpersonal disinhibition interspersed with episodes of rage in which she strikes out, spits, and swears.

The history was obtained from the Group Home Manager Jane; one of her one-to-one workers, Sylvia; as well as her mother, who participated in her assessment. The 5h and final neurosurgical intervention failed to control her seizures. Thus, two years ago she had vagal nerve stimulator implanted. This too failed to bring her seizures under control. By September this year she was having 50–60 seizures per day that typically clustered at night. In September she was started on Clobazam 20mg at night in addition to Carbamazepine CR 600mg twice a day. The addition of Clobazam has almost entirely eliminated the daytime seizures. At present, she is having 5–10 partial seizures a day. The seizure pattern is typical and is characterized by behavioural arrest followed by a stare and a myoclonic partial extension of her arms. She has an accompanying exclamation followed by a laugh. The entire episode lasts between 5–10 seconds. At no time has she ever lost postural control or dropped an item that she is holding. Two to three times a month she has a major motor convulsion in which each seizure lasts up to three minutes.

Behavioural control has improved significantly. Severe rage attacks occur approximately once every 6 weeks. Much of the behavioural control can be attributed to a tight behavioural program implemented by the personnel who work with her. They meet once a month to discuss behavioural strategies and harmonize their interventions. She can be talked down or can be distracted with a hug. Her workers attempt to anticipate difficult situations and avoid them.

Psychosocially she lives in a group home on her own supported by 24 hour one-to-one care. She has several activities in the community. On Monday she attends the library. On Tuesday she remains at home or goes to the gym. On Wednesday she attends Langley Lodge where she helps set up the lunch tables. On Thursday she helps deliver mail as well as attend Langley Lodge. On Friday she helps shred paper at a cooperative store for which she receives $5/week. She has breakfast on Saturday morning with her mother and workers. Sunday is devoted to cooking.

Her behaviour is intermittently characterized by stereotypic interpersonal interactions. Currently when excited she initiates a round of interaction with a statement "you won't belive it!"

The neurobehavioural inventory was completed by Ms. Jane Huff who manages the home. Suzanne eats independently and is continent of bowel and bladder. She is able to groom herself but needs help washing her hair, putting on her bra, or doing up her shoelaces and buttons. She is mobile and fully orientated. She is verbal, accessible, and spontaneously people seeking. She is capable of sustaining goal directed activities for 30–60 minutes. She has occasional motor restlessness characterized by pacing with occasional screaming. She remains interpersonally disinhibited characterized by irritable, loud, or silly behaviour and verbal and interpersonal intrusiveness. By contrast she is at times apathetic. Two to three times a week she will have an unpredictable aggressive outburst. This typically runs in cycles. She has some sexual disinhibition in which she will privately play with herself or display herself. She requires a moderate prompt to participate in activities of daily living but no prompts to comply with treatment.

Subjectively she had no complaints.

MENTAL STATUS EXAMINATION

She was alert and neatly groomed. Mental content was only partly accessible. This was likely due to compromised intellectual ability and poor cooperation. When stressed by a question she would distract herself by touching her worker Sylvia or getting up to address and touch her mother. She remained directable throughout with no aggressive outbursts notwithstanding the fact that some of the questions concerning her cognitive/intellectual functions placed her under significant stress leading to the displacement behaviour as described. Her speech was hyperprosodic and dysarthric. She was over familiar complementing me as well as hugging me as she left the office. Her mood was reported as euthymic. She disaffirmed any depression. Vegetative functions as per the collateral informants have been stable. None are aware of any significant depressive symptoms. Mental content was mostly inaccessible. She was able to provide a reasonable description of her weekly program. She disaffirmed any perceptual disturbances. This was also confirmed by her mother and caregivers.

Her higher intellectual functions were only superficially assessed. She was alert and orientated to person, day, month, year, date but not place. Digit span was 4 forwards. She was unable to perform digit span backwards because of an inability to comprehend the command. She recalled 0 out of 3 words following distraction. She recalled 3 out of 3 shapes following distraction. She had no difficulty copying the shapes from the three-word three shape test. However, she was unable to copy the three-dimensional cube. She doubled to 16.

A limited neurological examination was performed. She had a healed right craniotomy scar. She had a left hemianopia. Pupils were 4mm OU reacting to light both directly and consesually. She had a left central facial weakness. She had a moderate to severe hemiparesis. In the upper extremities she had a flicker of movement at the shoulder but 0 movements in the arm, forearm and hands. In the left leg she had 4/5 weakness of iliopsas, hamstrings and tibialis anterior. Tone of the left side was spastic. She had a left equinovarus posturing of the left foot with an Achilles tendon contracture. She had left sided pyer-reflexia. The sensory examination was unreliable. She walked with a marked hemiplegic gait.

ASSESSMENT

She continues to present with a disinhibited frontal lobe syndrome complicated by impaired cognitive/intellectual functions secondary to her right hemispherectomy and persistent seizures and characterized by social and interpersonal disinhibition and unpredictable rage attacks. The latter have significantly attenuated. This is almost certainly due to a consistent behavioural program that includes anticipation of difficult circumstances implemented by her care giving staff. Harmonized interventions have been greatly facilitated by their monthly meetings. Give her overall improvement she is not a candidate for medications that would dampen emotional reactivity. She is doing well with the current behavioural program which should be maintained. IF this deteriorates she would be a candidate for agents such as Nabilone. In the past she has be trialed on Clomipramine and Dextromorphan.

Seizure control remains an ongoing problem. This is being managed by Dr.Farrell. According to her mother he is contemplating a trial with Topiramate to help with weight loss.

I have not arranged for a follow-up appointment but advised that I would be willing to reassess on an annual basis or sooner should her behaviour deteriorate.

T.A.Hurwitz

APPENDIX C

Behavioural Supports Overview

BY CYNTHIA CLARK

When supporting Suzanna and her team to address her challenging behaviour, two basic assumptions inspired our philosophy:

The first assumption is that a person's behaviour, no matter how severe or bizarre, has meaning and serves a function for that person. It is the person's best attempt to communicate with others and the environment in order to get his/her needs met.

The second assumption is that we cannot understand what function the person's behaviour may be serving if we look only at the person. It is critical that we look at the person and the challenging behaviour(s) in the broad context in which the person is living and working.

Think About the Person:

- What do you know about the person's previous experiences (e.g. in a home, in an institution, with family, at a day program or workshop) that might help you understand him/her?
- Has the person done this behaviour in the past? If so, how might this behaviour have been adaptive for the person in the past? How did people respond and what was the outcome?
- When did the person's behaviour become a problem or begin to get worse?

- Have there been any recent changes in the person's life (e.g. staff leaving, holidays, family illness, death)?
- What do you know about the person's preferences, likes, dislikes, and fears?
- How many opportunities in his/her daily life does he/she have to engage in preferred activities?
- What are the person's skills overall? (e.g. What is he/she good at? In which areas may he/she need support?)
- Can the person effectively communicate (e.g. needs, wants, illness, emotions)?

Think About the General Social Climate:

- How much control do you think the person experiences in his/her daily life?
- Do those around the person provide ample opportunities for and encourage choice-making? Are those choices made by the person respected by others?
- What is the general quality of inter-personal interactions with others? (e.g. What is the ratio of "reprimands" to "praise" that the person receives from others?)
- Do those working with the person assist them in identifying and validating his/her feelings (e.g. through active listening)?

Think About the Environments:

- Where does the person live, work, and play? Do these environments meet the person's needs? Some questions to consider:

Home

- Does the person want to live there?
- Does the person like who they live with?
- Would the person prefer living with fewer or more persons?
- Does the person have things he/she likes to do?
- Does staffing allow the person the quantity and quality of attention needed?
- Does the location of and/or staffing allow the person adequate access to community recreational and leisure opportunities?
- Is the person involved in household routines?

Work

- Does the environment meet the person's needs (e.g. too large/too small; too noisy/too quiet; too much activity/not enough; type of co-workers; segregated/integrated)?
- What is the type of work available (e.g. varied/repetitive; outside/inside; physical/sedentary; individually/group; too difficult/too easy)?
- Is there a wage available? If so, consider:
    - Is it sufficient for needs?
    - Schedule of payment (e.g. monthly, weekly, daily)
    - Form of payment (e.g. cash, cheque)
    - Does the person make a connection between work, pay, and the acquisition of items/activities?

Caution must be exercised when the subject of problem behaviour is discussed. A specific problem behaviour is just one small example of a person's many behaviours. Staff must remember that every individual possesses desirable behaviours that are worthy of notice and reinforcement.

In thinking this way, staff will avoid the common practice of labelling an individual according to their problem behaviours. For example, some individuals have become known to staff only as "screamers" or "biters."

DEFINING A PROBLEM BEHAVIOUR

Behaviour is considered to be a problem if it has one or more of the following characteristics:

- is harmful or disruptive to oneself, others, or property (e.g. hitting oneself or others, writing on walls, yelling during meal time);
- interferes with learning new skills (e.g. moving fingers in front of one's eyes and staying in one's room for long periods of time);
- interferes with skills already learned (e.g. when the person is so involved in the behaviour that he/she no longer performs previously learned skills).

Problem behaviours are often not viewed as problems by a client. Problem behaviours usually have a payoff for the client and are more likely to be problems for peers and staff.

On the other hand, problem behaviours may not be problems for peers or staff, but are definitely harmful to the client (e.g. clients who sit and stare into space for long periods of time are engaging in behaviour that interfere with their learning).

UNDERSTANDING BEHAVIOUR
POINTS TO CONSIDER

- A person will continue a "negative behaviour" because, however unlikely this first appears, it is somehow adaptive. Our first task is to learn the answer to the question, "How is this adaptive?"
- Persons usually do things to get some beneficial result, not to get themselves labelled. If a person constantly seeks attention, he is

probably not looking for the label of "spoiled" or "manipulative" so much as a way of getting some needs fulfilled.

- The way that others behave and make us feel is probably the best clue to understanding how they themselves are feeling.
- The need for attention is a given. What is negotiable is how a person chooses to get it.
- Positive attention is harder to win than negative, but a person will choose to get negative attention rather than be ignored. The challenge is to help others understand that while positive social relationships are less predictable, they are, over time, more rewarding than negative relationships.
- Most persons tend to act the way they are expected to act.
- It is best not to ignore dramatic bids for attention. Ignoring only encourages a person to become more dramatic. It is better voluntarily to give attention to a person sooner than be forced into giving it later.
- When we pay attention to what a person does, we are not necessarily "reinforcing that behaviour." We are opening the opportunity for a dialogue with the person. When we ask, "What does this mean to you?" we are making the first step to answering the question, "How can we help?" The second step is to say, "If you need help with something, here is another way to ask for it." The message can be trusted only if we have demonstrated that we are worthy of trust.
- Persons seen as powerful have greater status and are more likely to be imitated. Persons who have a limited exposure to the ordinary range of powerful models in society will use the limited ones they know. Persons living in an institution will imitate those they understand to be powerful. Since one definition of power is the ability to decide for oneself, this sometimes leads to struggles when persons have decisions made for them.

- Power is inappropriately modelled when it is used to force. When we force someone to "cooperate," we are really announcing that power is the most important asset for being socially effective.
- Cooperation is a good model to show how we can take another's needs into consideration while at the same time taking care of our own.
- Persons who make us feel angry and hurt are expressing their own anger and hurt. If we respond with the sort of punishment we would find emotionally satisfying, then we are only continuing a cycle of frustration and pain.
- By isolating people's expression of their feelings, that is, by focusing on their "behaviour," we neglect the social system in which the behaviour takes place. Accordingly, we could eliminate a particular recurring example of something irritating that a person does, but we will not have satisfied the person's original wish.

SURVIVAL HINTS

- Planning sessions will likely be more successful when team members feel comfortable in giving their observations and suggestions **and** can also listen to other team members' observations, which may be different from their own.
- Once the information is collected and suggestions given, then the team needs to agree on a course of action or an intervention that all team members are willing to consistently follow.
- Team members need also to be committed to a process of periodic review and "fine tuning" of interventions in place.
- Keep your expectations realistic:
    - **Have a clear idea of what small changes need to happen that will indicate to you that the intervention is working.**
    - **Don't expect a 100 per cent success rate for any intervention.**

    - **Don't expect "overnight" change. Behavioural changes tend to happen slowly over time.**
    - **Don't be too discouraged if the first intervention is not successful. There are other options.**
- Whether an intervention is successful or not, it is always a learning process in which you continue to gather valuable information about the individual you are working with and about yourself.

I'm STILL The Bob and cathy's kid